Tucholsky Wagner Zola Scott Sydow Schlegel
Turgenev Wallace Fonatne Freud
Twain Walther von der Vogelweide Fouqué Friedrich II. von Preußen
Weber Freiligrath Frey
Fechner Weiße Rose von Fallersleben Kant Ernst Frommel
Fichte Hölderlin Richthofen
Engels Fielding Eichendorff Tacitus Dumas
Fehrs Faber Flaubert Eliasberg Ebner Eschenbach
Feuerbach Maximilian I. von Habsburg Fock Eliot Zweig Vergil
Goethe Ewald Elisabeth von Österreich London
Mendelssohn Balzac Shakespeare Dostojewski Ganghofer
Trackl Stevenson Lichtenberg Rathenau Doyle Gjellerup
Mommsen Tolstoi Lenz Hambruch Droste-Hülshoff
Thoma von Arnim Hanrieder
Dach Verne Hägele Hauff Humboldt
Karrillon Reuter Rousseau Hagen Hauptmann Gautier
Damaschke Defoe Hebbel Baudelaire
Descartes Hegel Kussmaul Herder
Wolfram von Eschenbach Darwin Dickens Schopenhauer Rilke George
Bronner Melville Grimm Jerome Bebel Proust
Campe Horváth Aristoteles Federer
Bismarck Vigny Barlach Voltaire Herodot
Gengenbach Heine
Storm Casanova Tersteegen Gilm Grillparzer Georgy
Brentano Chamberlain Lessing Langbein Gryphius
Strachwitz Claudius Schiller Lafontaine Kralik Iffland Sokrates
Katharina II. von Rußland Bellamy Schilling
Gerstäcker Raabe Gibbon Tschechow
Löns Hesse Hoffmann Gogol Wilde Gleim Vulpius
Luther Heym Hofmannsthal Klee Hölty Morgenstern Goedicke
Roth Heyse Klopstock Kleist
Luxemburg La Roche Puschkin Homer Mörike Musil
Machiavelli Horaz Kraft
Navarra Aurel Musset Kierkegaard Kraus
Nestroy Marie de France Lamprecht Kind Kirchhoff Hugo Moltke
Nietzsche Nansen Laotse Ipsen Liebknecht
Marx Lassalle Gorki Klett Leibniz Ringelnatz
von Ossietzky May vom Stein Lawrence Irving
Petalozzi Platon Knigge
Sachs Pückler Michelangelo Kafka
Poe Liebermann Kock
de Sade Praetorius Mistral Zetkin Korolenko

The publishing house tredition has created the series **TREDITION CLASSICS**. It contains classical literature works from over two thousand years. Most of these titles have been out of print and off the bookstore shelves for decades.

The book series is intended to preserve the cultural legacy and to promote the timeless works of classical literature. As a reader of a **TREDITION CLASSICS** book, the reader supports the mission to save many of the amazing works of world literature from oblivion.

The symbol of **TREDITION CLASSICS** is Johannes Gutenberg (1400 – 1468), the inventor of movable type printing.

With the series, tredition intends to make thousands of international literature classics available in printed format again – worldwide.

All books are available at book retailers worldwide in paperback and in hardcover. For more information please visit: www.tredition.com

tredition was established in 2006 by Sandra Latusseck and Soenke Schulz. Based in Hamburg, Germany, tredition offers publishing solutions to authors and publishing houses, combined with worldwide distribution of printed and digital book content. tredition is uniquely positioned to enable authors and publishing houses to create books on their own terms and without conventional manufacturing risks.

For more information please visit: www.tredition.com

The Cook's Decameron: a study in taste, containing over two hundred recipes for Italian dishes

W. G., Mrs. Waters

Imprint

This book is part of the TREDITION CLASSICS series.

Author: W. G., Mrs. Waters
Cover design: toepferschumann, Berlin (Germany)

Publisher: tredition GmbH, Hamburg (Germany)
ISBN: 978-3-8491-7088-2

www.tredition.com
www.tredition.de

Copyright:
The content of this book is sourced from the public domain.

The intention of the TREDITION CLASSICS series is to make world literature in the public domain available in printed format. Literary enthusiasts and organizations worldwide have scanned and digitally edited the original texts. tredition has subsequently formatted and redesigned the content into a modern reading layout. Therefore, we cannot guarantee the exact reproduction of the original format of a particular historic edition. Please also note that no modifications have been made to the spelling, therefore it may differ from the orthography used today.

To

A. V.

In memory of Certain Ausonian Feasts

Preface

Montaigne in one of his essays* mentions the high excellence Italian cookery had attained in his day. "I have entered into this Discourse upon the Occasion of an Italian I lately receiv'd into my Service, and who was Clerk of the Kitchen to the late Cardinal Caraffa till his Death. I put this Fellow upon an Account of his office: Where he fell to Discourse of this Palate-Science, with such a settled Countenance and Magisterial Gravity, as if he had been handling some profound Point of Divinity. He made a Learned Distinction of the several sorts of Appetites, of that of a Man before he begins to eat, and of those after the second and third Service: The Means simply to satisfy the first, and then to raise and acute the other two: The ordering of the Sauces, first in general, and then proceeded to the Qualities of the Ingredients, and their Effects: The Differences of Sallets, according to their seasons, which ought to be serv'd up hot, and which cold: The Manner of their Garnishment and Decoration, to render them yet more acceptable to the Eye after which he entered upon the Order of the whole Service, full of weighty and important Considerations."

It is consistent with Montaigne's large-minded habit thus to applaud the gifts of this master of his art who happened not to be a Frenchman. It is a canon of belief with the modern Englishman that the French alone can achieve excellence in the art of cookery, and when once a notion of this sort shall have found a lodgment in an Englishman's brain, the task of removing it will be a hard one. Not for a moment is it suggested that Englishmen or any one else should cease to recognise the sovereign merits of French cookery; all that is entreated is toleration, and perchance approval, of cookery of other schools. But the favourable consideration of any plea of this sort is hindered by the fact that the vast majority of Englishmen when they go abroad find no other school of cookery by the testing of which they may form a comparison. This universal prevalence of French cookery may be held to be a proof of its supreme excellence—that it is first, and the rest nowhere; but the victory is not so complete as it seems, and the facts would bring grief and humiliation rather than patriotic pride to the heart of a Frenchman like Brillat-Savarin. For the cookery we meet in the hotels of the great European cities, though it may be based on French traditions, is not the genuine thing, but a bastard, cosmopolitan growth, the same everywhere, and generally vapid and uninteresting. French cookery of the grand school suffers by being associated with such commonplace achievements. It is noted in the following pages how rarely English people on their travels penetrate where true Italian cookery may be tasted, wherefore it has seemed worth while to place within the reach of English housewives some Italian recipes which are especially fitted for the presentation of English fare to English palates under a different and not unappetising guise. Most of them will be found simple and inexpensive, and special care has been taken to include those recipes which enable the less esteemed portions of meat and the cheaper vegetables and fish to be treated more elaborately than they have hitherto been treated by English cooks.

The author wishes to tender her acknowledgments to her husband for certain suggestions and emendations made in the revision of the introduction, and for his courage in dining, "greatly daring," off many of the dishes. He still lives and thrives. Also to Mrs. Mitchell, her cook, for the interest and enthusiasm she has shown in the work, for her valuable advice, and for the care taken in testing the recipes.

CONTENTS

Preface

Prologue

PART I. THE COOK'S DECAMERON

The First Day

The Second Day

The Third Day

The Fourth Day

The Fifth Day

The Sixth Day

The Seventh Day

The Eighth Day

The Ninth Day

The Tenth Day

PART II — RECIPES

Sauces

No. 1. Espagnole, or Brown Sauce

No. 2. Velute Sauce

No. 3. Bechamel Sauce

No. 4. Mirepoix Sauce (for masking)

No. 5. Genoese Sauce

No. 6. Italian Sauce

No. 7. Ham Sauce, Salsa di Prosciutto

No. 8. Tarragon Sauce

No. 9. Tomato Sauce

No. 10. Tomato Sauce Piquante

No. 11. Mushroom Sauce

No. 12. Neapolitan Sauce

No. 13. Neapolitan Anchovy Sauce

No. 14. Roman Sauce (Salsa Agro-dolce)

No. 15. Roman Sauce (another way)

No. 16. Supreme Sauce

No. 17. Pasta marinate (For masking Italian Frys)

No. 18. White Villeroy

Soups

No. 19. Clear Soup

No. 20. Zuppa Primaverile (Spring Soup)

No. 21. Soup alla Lombarda

No. 22. Tuscan Soup

No. 23. Venetian Soup

No. 24. Roman Soup

No. 25. Soup alla Nazionale

No. 26. Soup alla Modanese

No. 27. Crotopo Soup

No. 28. Soup all'Imperatrice

No. 29. Neapolitan Soup

No. 30. Soup with Risotto

No. 31. Soup alla Canavese

No. 32. Soup alla Maria Pia

No. 33. Zuppa d' Erbe (Lettuce Soup)

No. 34. Zuppa Regina di Riso (Queen's Soup)

Minestre

No. 35. A Condiment for Seasoning Minestre, &c.

No. 36. Minestra alla Casalinga

No. 37. Minestra of Rice and Turnips

No. 38. Minestra alla Capucina

No. 39. Minestra of Semolina

No. 40. Minestrone alla Milanese

No. 41. Minestra of Rice and Cabbage

No. 42. Minestra of Rice and Celery

Fish

No. 43. Anguilla alla Milanese (Eels).

No. 44. Filletti di Pesce alla Villeroy (Fillets of Fish)

No. 45. Astachi all'Italiana (Lobster)

No. 46. Baccala alla Giardiniera (Cod)

No. 47. Triglie alla Marinara (Mullet)

No. 48. Mullet alla Tolosa

No. 49. Mullet alla Triestina

No. 50. Whiting alla Genovese

No. 51. Merluzzo in Bianco (Cod)

No. 52. Merluzzo in Salamoia (Cod)

No. 53. Baccala in Istufato (Haddock)

No. 54. Naselli con Piselli (Whiting)

No. 55. Ostriche alla Livornese (Oysters)

No. 56. Ostriche alla Napolitana (Oysters)

No. 57. Ostriche alla Veneziana (Oysters)

No. 58. Pesci diversi alla Casalinga (Fish)

No. 59. Pesce alla Genovese (Sole or Turbot)

No. 60. Sogliole in Zimino (Sole)

No. 61. Sogliole al tegame (Sole)

No. 62. Sogliole alla Livornese (Sole)

No. 63. Sogliole alla Veneziana (Sole)

No. 64. Sogliole alla Parmigiana (Sole).*

No. 65. Salmone alla Genovese (Salmon)

No. 66. Salmone alla Perigo (Salmon)

No. 67. Salmone alla giardiniera (Salmon)

No. 68. Salmone alla Farnese (Salmon)

No. 69. Salmone alla Santa Fiorentina (Salmon)

No. 70. Salmone alla Francesca (Salmon)

No. 71. Fillets of Salmon in Papiliotte

Beef, Mutton, Veal, Lamb, &C.

No. 72. Manzo alla Certosina (Fillet of Beef)

No. 73. Stufato alla Florentina (Stewed Beef)

No. 74. Coscia di Manzo al Forno (Rump Steak)

No. 75. Polpettine alla Salsa Piccante (Beef Olives)

No. 76. Stufato alla Milanese (Stewed Beef)

No. 77. Manzo Marinato Arrosto (Marinated Beef)

No. 78. Manzo con sugo di Barbabietole (Fillet of Beef)

No. 79. Manzo in Insalata (Marinated Beef)

No. 80. Filetto di Bue con Pistacchi (Fillets of Beef with Pistacchios)

No. 81. Scalopini di Riso (Beef with Risotto)

No. 82. Tenerumi alla Piemontese (Tendons of Veal)

No. 83. Bragiuole di Vitello (Veal Cutlets)

No. 84. Costolette alla Manza (Veal Cutlets)

No. 85. Vitello alla Pellegrina (Breast of Veal)

No. 86. Frittura Piccata al Marsala (Fillet of Veal)

No. 87. Polpettine Distese (Veal Olives)

No. 88. Coste di Vitello Imboracciate (Ribs of Veal)

No. 89. Costolette di Montone alla Nizzarda (Mutton Cutlets)

No. 90. Petto di Castrato all'Italiana (Breast of Mutton)

No. 91. Petto di Castrato alla Salsa piccante (Breast of Mutton)

No. 92. Tenerumi d'Agnello alla Villeroy (Tendons of Lamb)

No. 93. Tenerumi d' Agnello alla Veneziana (Tendons of Lamb)

No. 94. Costolette d' Agnello alla Costanza (Lamb Cutlets)

Tongue, Sweetbread, Calf's Head, Liver, Sucking Pig, &C.

No. 95. Timballo alla Romana

No. 96. Timballo alla Lombarda

No. 97. Lingua alla Visconti (Tongue)

No. 98. Lingua di Manzo al Citriuoli (Tongue with Cucumber)

No. 99. Lingue di Castrato alla Cuciniera (Sheep's Tongues)

No. 100. Lingue di Vitello all'Italiana (Calves' Tongues)

No. 101. Porcelletto alla Corradino (Sucking Pig)

No. 102. Porcelletto da Latte in Galantina (Sucking Pig)

No. 103. Ateletti alla Sarda

No. 104. Ateletti alla Genovese

No. 105. Testa di Vitello alla Sorrentina (Calf's Head)

No. 106. Testa di Vitello con Salsa Napoletana (Calf's Head)

No. 107. Testa di Vitello alla Pompadour (Calf's Head)

No. 108. Testa di Vitello alla Sanseverino (Calf's Head)

No. 109. Testa di Vitello in Frittata (Calf's Head)

No. 110. Zampetti (Calves' Feet)

No. 111. Bodini Marinati

No. 112. Animelle alla Parmegiana (Sweetbread)

No. 113. Animelle in Cartoccio (Sweetbread)

No. 114. Animelle all'Italiana (Sweetbread)

No. 115. Animelle Lardellate (Sweetbread)

No. 116. Frittura di Bottoni e di Animelle (Sweetbread and Mushrooms)

No. 117. Cervello in Fili serbe (Calf's Brains)

No. 118. Cervello alla Milanese (Calf's Brains)

No. 119. Cervello alla Villeroy (Calf's Brains)

No. 120. Frittura of Liver and Brains

No. 121. Cervello in Frittata Montano (Calf's Brains)

No. 122. Marinata di Cervello alla Villeroy (Calf's Brains)

No. 123. Minuta alla Milanese (Lamb's Sweetbread)

No. 124. Animelle al Sapor di Targone (Lamb's Fry)

No. 125. Fritto Misto alla Villeroy

No. 126. Fritto Misto alla Piemontese

No. 127. Minuta di Fegatini (Ragout of Fowls' Livers)

No. 128. Minuta alla Visconti (Chickens' Livers)

No. 129. Croutons alla Principesca

No. 130. Croutons alla Romana

Fowl, Duck, Game, Hare, Rabbit, &c.

No. 131. Soffiato di Cappone (Fowl Souffle)

No. 132. Pollo alla Fiorentina (Chicken)

No. 133. Pollo all'Oliva (Chicken)

No. 134. Pollo alla Villereccia (Chicken)

No. 135. Pollo alla Cacciatora (Chicken)

No. 136. Pollastro alla Lorenese (Fowl)

No. 137. Pollastro in Fricassea al Burro (Fowl)

No. 138. Pollastro in istufa di Pomidoro (Braized Fowl)

No. 139. Cappone con Riso (Capon with Rice)

No. 140. Dindo Arrosto alla Milanese (Roast Turkey)

No. 141. Tacchinotto all'Istrione (Turkey Poult)

No. 142. Fagiano alla Napoletana (Pheasant)

No. 143. Fagiano alla Perigo (Pheasant)

No. 144. Anitra Selvatica (Wild Duck)

No. 145. Perniciotti alla Gastalda (Partridges)

No. 146. Beccaccini alla Diplomatica (Snipe)

No. 147. Piccioni alla minute (Pigeons)

No. 148. Piccioni in Ripieno (Stuffed Pigeons)

No. 149. Lepre in istufato (Stewed Hare)

No. 150. Lepre Agro-dolce (Hare)

No. 151. Coniglio alla Provenzale (Rabbit)

No. 152. Coniglio arrostito alla Corradino (Roast Rabbit)

No. 153. Coniglio in salsa Piccante (Rabbit)

Vegetables

No. 154. Asparagi alla salsa Suprema (Asparagus)

No. 155. Cavoli di Bruxelles alla Savoiarda (Brussels Sprouts)

No. 156. Barbabietola alla Parmigiana (Beetroot)

No. 157. Fave alla Savoiarda (Beans)

No. 158. Verze alla Capuccina (Cabbage)

No. 159. Cavoli fiodi alla Lionese (Cauliflower)

No. 160. Cavoli fiodi fritti (Cauliflower)

No. 161. Cauliflower alla Parmigiana

No. 162. Cavoli Fiori Ripieni

No. 163. Sedani alla Parmigiana (Celery)

No. 164. Sedani fritti all'Italiana (Celery)

No. 165. Cetriuoli alla Parmigiana (Cucumber)

No. 166. Cetriuoli alla Borghese (Cucumber)

No. 167. Carote al sughillo (Carrots)

No. 168. Carote e piselli alla panna (Carrots and Peas)

No. 169. Verze alla Certosine (Cabbage)

No. 170. Lattughe al sugo (Lettuce)

No. 171 Lattughe farcite alla Genovese (Lettuce)

No. 172. Funghi cappelle infarcite (Stuffed Mushrooms)

No. 173. Verdure miste (Macedoine of Vegetables)

No. 174. Patate alla crema (Potatoes in cream)

No. 175. Cestelline di patate alla giardiniera (Potatoes)

No. 176. Patate al Pomidoro (Potatoes with Tomato Sauce)

No. 177. Spinaci alla Milanese (Spinach)

No. 178. Insalata di patate (Potato salad)

No. 179. Insalata alla Navarino (Salad)

No. 180. Insalata di pomidoro (Tomato Salad)

No. 181. Tartufi alla Dino (Truffles)

Macaroni, Rice, Polenta, and Other Italian Pastes

No. 182. Macaroni with Tomatoes

No. 183. Macaroni alla Casalinga

No. 184. Macaroni al Sughillo

No. 185. Macaroni alla Livornese

No. 186. Tagliarelle and Lobster

No. 187. Polenta

No. 188. Polenta Pasticciata

No. 189. Battuffoli

No. 190. Risotto all'Italiana

No. 191. Risotto alla Genovese

No. 192. Risotto alla Spagnuola

No. 193. Risotto alla Capuccina

No. 194. Risotto alla Parigina

No. 195. Ravioli

No. 196. Ravioli alla Fiorentina

No. 197. Gnocchi alla Romana

No. 198. Gnocchi alla Lombarda

No. 199. Frittata di Riso (Savoury Rice Pancake)

Omelettes And Other Egg Dishes

No. 200. Uova al Tartufi (Eggs with Truffles)

No. 201. Uova al Pomidoro (Eggs and Tomatoes)

No. 202. Uova ripiene (Canapes of Egg)

No. 203. Uova alla Fiorentina (Eggs)

No. 204. Uova in fili (Egg Canapes)

No. 205. Frittata di funghi (Mushroom Omelette)

No. 206. Frittata con Pomidoro (Tomato Omelette)

No. 207. Frittata con Asparagi (Asparagus Omelette)

No. 208. Frittata con erbe (Omelette with Herbs)

No. 209. Frittata Montata (Omelette Souffle)

No. 210. Frittata di Prosciutto (Ham Omelette)

Sweets and Cakes

No. 211. Bodino of Semolina

No. 212. Crema rappresa (Coffee Cream)

No. 213. Crema Montata alle Fragole (Strawberry Cream)

No. 214. Croccante di Mandorle (Cream Nougat)

No. 215. Crema tartara alla Caramella (Caramel Cream)

No. 216. Cremona Cake

No. 217. Cake alla Tolentina

No. 218. Riso all'Imperatrice

No. 219. Amaretti leggieri (Almond Cakes)

No. 220. Cakes alla Livornese

No. 221. Genoese Pastry

No. 222. Zabajone

No. 223. Iced Zabajone

No. 224. Pan-forte di Siena (Sienese Hardbake)

New Century Sauce

No. 225. Fish Sauce

No. 226. Sauce Piquante (for Meat, Fowl, Game, Rabbit, &c.)

No. 227. Sauce for Venison, Hare, &c.

No. 228. Tomato Sauce Piquante

No. 229. Sauce for Roast Pork, Ham, &c.

No. 230. For masking Cutlets, &c.

PART I. THE COOK'S DECAMERON

Prologue

The Marchesa di Sant'Andrea finished her early morning cup of tea, and then took up the batch of correspondence which her maid had placed on the tray. The world had a way of treating her in kindly fashion, and hostile or troublesome letters rarely veiled their ugly faces under the envelopes addressed to her; wherefore the perfection of that pleasant half-hour lying between the last sip of tea and the first step to meet the new day was seldom marred by the perusal of her morning budget. The apartment which she graced with her seemly presence was a choice one in the Mayfair Hotel, one which she had occupied for the past four or five years during her spring visit to London; a visit undertaken to keep alive a number of pleasant English friendships which had begun in Rome or Malta. London had for her the peculiar attraction it has for so many Italians, and the weeks she spent upon its stones were commonly the happiest of the year.

The review she took of her letters before breaking the seals first puzzled her, and then roused certain misgivings in her heart. She recognised the handwriting of each of the nine addresses, and at the same time recalled the fact that she was engaged to dine with every one of the correspondents of this particular morning. Why should they all be writing to her? She had uneasy forebodings of postponement, and she hated to have her engagements disturbed; but it was useless to prolong suspense, so she began by opening the envelope addressed in the familiar handwriting of Sir John Oglethorpe, and this was what Sir John had to say —

"My Dear Marchesa, words, whether written or spoken, are powerless to express my present state of mind. In the first place, our dinner on Thursday is impossible, and in the second, I have lost Narcisse and forever. You commented favourably upon that supreme of lobster and the Ris de Veau a la Renaissance we tasted last week, but never again will you meet the handiwork of Narcisse. He came to me with admirable testimonials as to his artistic excellence; with regard to his moral past I was, I fear, culpably negligent, for I now learn that all the time he presided over my stewpans he was wanted by the French police on a charge of murdering his wife. A young lady seems to have helped him; so I fear Narcisse has broken more than one of the commandments in this final escapade. The truly great have ever been subject to these momentary aberrations, and Narcisse being now in the hands of justice—so called—our dinner must needs stand over, though not, I hope, for long. Meantime the only consolation I can perceive is the chance of a cup of tea with you this afternoon."

"J. O."

Sir John Oglethorpe had been her husband's oldest and best friend. He and the Marchesa had first met in Sardinia, where they had both of them gone in pursuit of woodcock, and since the Marchesa had been a widow, she and Sir John had met either in Rome or in London every year. The dinner so tragically manqué had been arranged to assemble a number of Anglo-Italian friends; and, as Sir John was as perfect as a host as Narcisse was as a cook, the disappointment was a heavy one. She threw aside the letter with a gesture of vexation, and opened the next.

"Sweetest Marchesa," it began, "how can I tell you my grief at having to postpone our dinner for Friday. My wretched cook (I gave her seventy-five pounds a year), whom I have long suspected of intemperate habits, was hopelessly inebriated last night, and had to be conveyed out of the house by my husband and a dear, devoted friend who happened to be dining with us, and deposited in a four-wheeler. May I look in tomorrow afternoon and pour out my grief to you? Yours cordially,

"Pamela St. Aubyn Fothergill."

When the Marchesa had opened four more letters, one from Lady Considine, one from Mrs. Sinclair, one from Miss Macdonnell, and one from Mrs. Wilding, and found that all these ladies were obliged to postpone their dinners on account of the misdeeds of their cooks, she felt that the laws of average were all adrift. Surely the three remaining letters must contain news of a character to counterbalance what had already been revealed, but the event showed that, on this particular morning, Fortune was in a mood to strike hard. Colonel Trestrail, who gave in his chambers carefully devised banquets, compounded by a Bengali who was undoubtedly something of a genius, wrote to say that this personage had left at a day's notice, in order to embrace Christianity and marry a lady's-maid who had just come into a legacy of a thousand pounds under the will of her late mistress. Another correspondent, Mrs. Gradinger, wrote that her German cook had announced that the dignity of womanhood was, in her opinion, slighted by the obligation to prepare food for others in exchange for mere pecuniary compensation. Only on condition of the grant of perfect social equality would she consent to stay, and Mrs. Gradinger, though she held advanced opinions, was hardly advanced far enough to accept this suggestion. Last of all, Mr. Sebastian van der Roet was desolate to announce that his cook, a Japanese, whose dishes were, in his employer's estimation, absolute inspirations, had decamped and taken with him everything of value he could lay hold of; and more than desolate, that he was forced to postpone the pleasure of welcoming the Marchesa di Sant' Andrea at his table.

When she had finished reading this last note, the Marchesa gathered the whole mass of her morning's correspondence together, and uttering a few Italian words which need not be translated, rolled it into a ball and hurled the same to the farthest corner of the room. "How is it," she ejaculated, "that these English, who dominate the world abroad, cannot get their food properly cooked at home? I suppose it is because they, in their lofty way, look upon cookery as a non-essential, and consequently fall victims to gout and dyspepsia, or into the clutches of some international brigandaccio, who declares he is a cordon bleu. One hears now and again pleasant remarks about the worn-out Latin races, but I know of one Latin race which can do better than this in cookery." And having thus delivered herself, the Marchesa lay back on the pillows and reviewed the situation.

She was sorry in a way to miss the Colonel's dinner. The dishes which the Bengali cook turned out were excellent, but the host himself was a trifle dictatorial and too fond of the sound of his own voice, while certain of the inevitable guests were still worse. Mrs. Gradinger's letter came as a relief; indeed the Marchesa had been wondering why she had ever consented to go and pretend to enjoy herself by eating an ill-cooked dinner in company with social reformers and educational prigs. She really went because she liked Mr. Gradinger, who was as unlike his wife as possible, a stout youth of forty, with a breezy manner and a decided fondness for sport. Lady Considine's dinners were indifferent, and the guests were apt to be a bit too smart and too redolent of last season's Monte Carlo odour. The Sinclairs gave good dinners to perfectly selected guests, and by reason of this virtue, one not too common, the host and hostess might be pardoned for being a little too well satisfied with themselves and with their last new bibelot. The Fothergill dinners were like all other dinners given by the Fothergills of society. They were costly, utterly undistinguished, and invariably graced by the presence of certain guests who seemed to have been called in out of the street at the last moment. Van der Roet's Japanese menus were curious, and at times inimical to digestion, but the personality of the host was charming. As to Sir John Oglethorpe, the question of the dinner postponed troubled her little: another repast, the finest that London's finest restaurant could furnish, would certainly be forthcoming before long. In Sir John's case, her discomposure took the form of sympathy for her friend in his recent bereavement. He had been searching all his life for a perfect cook, and he had found, or believed he had found, such an one in Narcisse; wherefore the Marchesa was fully persuaded that, if that artist should evade the guillotine, she would again taste his incomparable handiwork, even though he were suspected of murdering his whole family as well as the partner of his joys.

That same afternoon a number of the balked entertainers foregathered in the Marchesa's drawing-room, the dominant subject of discourse being the approaching dissolution of London society from the refusal of one human to cook food for another. Those present were gathered in two groups. In one the Colonel, in spite of the recent desertion of his Oriental, was asserting that the Government should be required to bring over consignments of perfectly trained Indian cooks, and thus trim the balance between dining room and kitchen; and to the other Mrs. Gradinger, a gaunt, ill-dressed lady in spectacles, with a commanding nose and dull, wispy hair, was proclaiming in a steady metallic voice, that it was absolutely necessary to double the school rate at once in order to convert all the girls and some of the boys as well, into perfectly equipped food-cooking animals; but her audience gradually fell away, and in an interval of silence the voice of the hostess was heard giving utterance to a tentative suggestion.

"But, my dear, it is inconceivable that the comfort and the movement of society should depend on the humours of its servants. I don't blame them for refusing to cook if they dislike cooking, and can find other work as light and as well paid; but, things being as they are, I would suggest that we set to work somehow to make ourselves independent of cooks."

"That 'somehow' is the crux, my dear Livia," said Mrs. Sinclair. "I have a plan of my own, but I dare not breathe it, for I'm sure Mrs. Gradinger would call it 'anti-social,' whatever that may mean."

"I should imagine that it is a term which might be applied to any scheme which robs society of the ministrations of its cooks," said Sir John.

"I have heard mathematicians declare that what is true of the whole is true of its parts," said the Marchesa. "I daresay it is, but I never stopped to inquire. I will amplify on my own account, and lay down that what is true of the parts must be true of the whole. I'm sure that sounds quite right. Now I, as a unit of society, am independent of cooks because I can cook myself, and if all the other units were independent, society itself would be independent—ecco!"

"To speak in this tone of a serious science like Euclid seems rather frivolous," said Mrs. Gradinger. "I may observe—" but here mercifully the observation was checked by the entry of Mrs. St. Aubyn Fothergill.

She was a handsome woman, always dominated by an air of serious preoccupation, sumptuously, but not tastefully dressed. In the social struggle upwards, wealth was the only weapon she possessed, and wealth without dexterity has been known to fail before this. She made efforts, indeed, to imitate Mrs. Sinclair in the elegancies of menage, and to pose as a woman of mind after the pattern of

Mrs. Gradinger; but the task first named required too much tact, and the other powers of endurance which she did not possess.

"You'll have some tea, Mrs. Fothergill?" said the Marchesa. "It's so good of you to have come."

"No, really, I can't take any tea; in fact, I couldn't take any lunch out of vexation at having to put you off, my dear Marchesa."

"Oh, these accidents will occur. We were just discussing the best way of getting round them," said the Marchesa. "Now, dear," — speaking to Mrs. Sinclair — "let's have your plan. Mrs. Gradinger has fastened like a leech on the Canon and Mrs. Wilding, and won't hear a word of what you have to say."

"Well, my scheme is just an amplification of your mathematical illustrations, that we should all learn to cook for ourselves. I regard it no longer as impossible, or even difficult, since you have informed us that you are a mistress of the art. We'll start a new school of cookery, and you shall teach us all you know."

"Ah, my dear Laura, you are like certain English women in the hunting field. You are inclined to rush your fences," said the Marchesa with a deprecatory gesture. "And just look at the people gathered here in this room. Wouldn't they — to continue the horsey metaphor — be rather an awkward team to drive?"

"Not at all, if you had them in suitable surroundings. Now, supposing some beneficent millionaire were to lend us for a month or so a nice country house, we might install you there as Mistress of the stewpans, and sit at your feet as disciples," said Mrs. Sinclair.

"The idea seems first-rate," said Van der Roet; "and I suppose, if we are good little boys and girls, and learn our lessons properly, we may be allowed to taste some of our own dishes."

"Might not that lead to a confusion between rewards and punishments?" said Sir John.

"If ever it comes to that," said Miss Macdonnell with a mischievous glance out of a pair of dark, flashing Celtic eyes, "I hope that our mistress will inspect carefully all pupils' work before we are asked to eat it. I don't want to sit down to another of Mr. Van der Roet's Japanese salads made of periwinkles and wallflowers."

"And we must first catch our millionaire," said the Colonel.

During these remarks Mrs. Fothergill had been standing "with parted lips and straining eyes," the eyes of one who is seeking to "cut in." Now came her chance. "What a delightful idea dear Mrs. Sinclair's is. We have been dreadfully extravagant this year over buying pictures, and have doubled our charitable subscriptions, but I believe I can still promise to act in a humble way the part of Mrs. Sinclair's millionaire. We have just finished doing up the 'Laurestinas,' a little place we bought last year, and it is quite at your service, Marchesa, as soon as you liketo occupy it."

This unlooked-for proposition almost took away the Marchesa's breath. "Ah, Mrs. Fothergill," she said, "it was Mrs. Sinclair's plan, not mine. She kindly wishes to turn me into a cook for I know not how long, just at the hottest season of the year, a fate I should hardly have chosen for myself."

"My dear, it would be a new sensation, and one you would enjoy beyond everything. I am sure it is a scheme every one here will hail with acclamation," said Mrs. Sinclair. All other conversation had now ceased, and the eyes of the rest of the company were fixed on the speaker. "Ladies and gentlemen," she went on, "you have heard my suggestion, and you have heard Mrs. Fothergill's most kind and opportune offer of her country house as the seat of our school of cookery. Such an opportunity is one in ten thousand. Surely all of us—-even the Marchesa—must see that it is one not to be neglected."

"I approve thoroughly," said Mrs. Gradinger; "the acquisition of knowledge, even in so material a field as that of cookery, is always a clear gain."

"It will give Gradinger a chance to put in a couple of days at Ascot," whispered Van der Roet.

"Where Mrs. Gradinger leads, all must follow," said Miss Macdonnell. "Take the sense of the meeting, Mrs. Sinclair, before the Marchesa has time to enter a protest."

"And is the proposed instructress to have no voice in the matter?" said the Marchesa, laughing.

"None at all, except to consent," said Mrs. Sinclair; "you are going to be absolute mistress over us for the next fortnight, so you surely might obey just this once."

"You have been denouncing one of our cherished institutions, Marchesa," said Lady Considine, "so I consider you are bound to help us to replace the British cook by something better."

"If Mrs. Sinclair has set her heart on this interesting experiment. You may as well consent at once, Marchesa," said the Colonel, "and teach us how to cook, and—what may be a harder task—to teach us to eat what other aspirants may have cooked."

"If this scheme really comes off," said Sir John, "I would suggest that the Marchesa should always be provided with a plate of her own up her sleeve—if I may use such an expression—so that any void in the menu, caused by failure on the part of the under-skilled or over-ambitious amateur, may be filled by what will certainly be a chef-d'oeuvre."

"I shall back up Mrs. Sinclair's proposition with all my power," said Mrs. Wilding. "The Canon will be in residence at Martlebridge for the next month, and I would much rather be learning cookery under the Marchesa than staying with my brother-in-law at Ealing."

"You'll have to do it, Marchesa," said Van der Roet; "when a new idea catches on like this, there's no resisting it."

"Well, I consent on one condition—that my rule shall be absolute," said the Marchesa, "and I begin my career as an autocrat by giving Mrs. Fothergill a list of the educational machinery I shall want, and commanding her to have them all ready by Tuesday morning, the day on which I declare the school open."

A chorus of applause went up as soon as the Marchesa ceased speaking.

"Everything shall be ready," said Mrs. Fothergill, radiant with delight that her offer had been accepted, "and I will put in a full staff of servants selected from our three other establishments."

"Would it not be as well to send the cook home for a holiday?" said the Colonel. "It might be safer, and lead to less broth being spoilt."

"It seems," said Sir John, "that we shall be ten in number, and I would therefore propose that, after an illustrious precedent, we limit our operations to ten days. Then if we each produce one culinary poem a day we shall, at the end of our time, have provided the world with a hundred new reasons for enjoying life, supposing, of course, that we have no failures. I propose, therefore, that our society be called the 'New Decameron.'"

"Most appropriate," said Miss Macdonnell, "especially as it owes its origin to an outbreak of plague — the plague in the kitchen."

The First Day

On the Tuesday morning the Marchesa travelled down to the "Laurestinas," where she found that Mrs. Fothergill had been as good as her word. Everything was in perfect order. The Marchesa had notified to her pupils that they must report themselves that same evening at dinner, and she took down with her her maid, one of those marvellous Italian servants who combine fidelity with efficiency in a degree strange to the denizens of more progressive lands. Now, with Angelina's assistance, she proposed to set before the company their first dinner all'Italiana, and the last they would taste without having participated in the preparation. The real work was to begin the following morning.

The dinner was both a revelation and a surprise to the majority of the company. All were well travelled, and all had eaten of the mongrel French dishes given at the "Grand" hotels of the principal Italian cities, and some of them, in search of adventures, had dined at London restaurants with Italian names over the doors, where — with certain honourable exceptions — the cookery was French, and not of the best, certain Italian plates being included in the carte for a regular clientele, dishes which would always be passed over by the English investigator, because he now read, or tried to read, their names for the first time. Few of the Marchesa's pupils had ever wandered away from the arid table d'hote in Milan, or Florence, or Rome, in search of the ristorante at which the better class of townsfolk were wont to take their colazione. Indeed, whenever an Englishman does break fresh ground in this direction, he rarely finds sufficient presence of mind to controvert the suggestions of the smiling minister who, having spotted his Inglese, at once marks down an omelette aux fines herbes and a biftek aux pommes as the only food such a creature can consume. Thus the culinary experiences of Englishmen in Italy have led to the perpetuation of the legend that the traveller can indeed find decent food in the large towns, "because the cooking there is all French, you know," but that, if he should deviate from the beaten track, unutterable horrors, swimming in oil and reeking with garlic, would be his portion. Oil and garlic are in popular English belief the inseparable accidents of Italian cookery, which is supposed to gather its solitary claim to individuality from the never-failing presence of these admirable, but easily abused, gifts of Nature.

"You have given us a delicious dinner, Marchesa," said Mrs. Wilding as the coffee appeared. "You mustn't think me captious in my remarks—indeed it would be most ungracious to look a gift-dinner in the—What are you laughing at, Sir John? I suppose I've done something awful with my metaphors—mixed them up somehow."

"Everything Mrs. Wilding mixes will be mixed admirably, as admirably, say, as that sauce which was served with the Manzo alla Certosina," Sir John replied.

"That is said in your best style, Sir John," replied Mrs. Wilding; "but what I was going to remark was, that I, as a poor parson's wife, shall ask for some instruction in inexpensive cooking before we separate. The dinner we have just eaten is surely only within the reach of rich people."

"I wish some of the rich people I dine with could manage now and then to reach a dinner as good," said the Colonel.

"I believe it is a generally received maxim, that if you want a truth to be accepted you must repeat the same in season and out, whenever you have the opportunity," said the Marchesa. "The particular truth I have now in mind is the fact that Italian cookery is the cookery of a poor nation, of people who have scant means wherewith to purchase the very inferior materials they must needs work with;

and that they produce palatable food at all is, I maintain, a proof that they bring high intelligence to the task. Italian culinary methods have been developed in the struggle when the cook, working with an allowance upon which an English cook would resign at once, has succeeded by careful manipulation and the study of flavouring in turning out excellent dishes made of fish and meat confessedly inferior. Now, if we loosen the purse-strings a little, and use the best English materials, I affirm that we shall achieve a result excellent enough to prove that Italian cookery is worthy to take its stand beside its great French rival. I am glad Mrs. Wilding has given me an opportunity to impress upon you all that its main characteristics are simplicity and cheapness, and I can assure her that, even if she should reproduce the most costly dishes of our course, she will not find any serious increase in her weekly bills. When I use the word simplicity, I allude, of course, to everyday cooking. Dishes of luxury in any school require elaboration, care, and watchfulness."

> Menu—Dinner {*} Zuppa d'uova alla Toscana. Tuscan egg-soup. Sogliole alla Livornese. Sole alla Livornese. Manzo alla Certosina. Fillet of beef, Certosina sauce. Minuta alla Milanese. Chickens' livers alla Milanese. Cavoli fiodi ripieni. Cauliflower with forcemeat. Cappone arrosto con insalata. Roast capon with salad. Zabajone. Spiced custard. Uova al pomidoro. Eggs and tomatoes.

* The recipes for the dishes contained in all these menus will be found in the second part of the book. The limits of the seasons have necessarily been ignored.

The Second Day

Wednesday's luncheon was anticipated with some curiosity, or even searchings of heart, as in it would appear the first-fruits of the hand of the amateur. The Marchesa wisely restricted it to two dishes, for the compounding of which she requisitioned the services of

Lady Considine, Mrs. Sinclair, and the Colonel. The others she sent to watch Angelina and her circle while they were preparing the vegetables and the dinner entrees. After the luncheon dishes had been discussed, they were both proclaimed admirable. It was a true bit of Italian finesse on the part of the Marchesa to lay a share of the responsibility of the first meal upon the Colonel, who was notoriously the most captious and the hardest to please of all the company; and she did even more than make him jointly responsible, for she authorised him to see to the production of a special curry of his own invention, the recipe for which he always carried in his pocketbook, thus letting India share with Italy in the honours of the first luncheon.

"My congratulations to you on your curry, Colonel Trestrail," said Miss Macdonnell. "You haven't followed the English fashion of flavouring a curry by emptying the pepper-pot into the dish?"

"Pepper properly used is the most admirable of condiments," the Colonel said.

"Why this association of the Colonel and pepper?" said Van der Roet. "In this society we ought to be as nice in our phraseology as in our flavourings, and be careful to eschew the incongruous. You are coughing, Mrs. Wilding. Let me give you some water."

"I think it must have been one of those rare grains of the Colonel's pepper, for you must have a little pepper in a curry, mustn't you, Colonel? Though, as Miss Macdonnell says, English cooks generally overdo it."

"Vander is in one of his pleasant witty moods," said the Colonel, "but I fancy I know as much about the use of pepper as he does about the use of oil colours; and now we have, got upon art criticism, I may remark, my dear Vander, I have been reminded that you have been poaching on my ground. I saw a landscape of yours the other day, which looked as if some of my curry powder had got into the sunset. I mean the one poor blind old Wilkins bought at your last show."

"Ah, but that sunset was an inspiration, Colonel, and consequently beyond your comprehension."

"It is easy to talk of inspiration," said Sir John, "and, perhaps, now that we are debating a matter of real importance, we might spend our time more profitably than in discussing what is and what is not a good picture. Some inspiration has been brought into our symposium, I venture to affirm that the brain which devised and the hand which executed the Tenerumi di Vitello we have just tasted, were both of them inspired. In the construction of this dish there is to be recognised a breath of the same afflatus which gave us the Florentine campanile, and the Medici tombs, and the portrait of Monna Lisa. When we stand before any one of these masterpieces, we realise at a glance how keen must have been the primal insight, and how strenuous the effort necessary for the evolution of so consummate an achievement; and, with the savour of the Tenerumi di Vitello still fresh, I feel that it deserves to be added to the list of Italian capo lavori. Now, as I was not fortunate enough to be included in the pupils' class this morning, I must beg the next time the dish is presented to us—and I imagine all present will hail its renaissance with joy—that I may be allowed to lend a hand, or even a finger, in its preparation."

"Veal, with the possible exception of Lombard beef, is the best meat we get in Italy," said the Marchesa, "so an Italian cook, when he wants to produce a meat dish of the highest excellence, generally turns to veal as a basis. I must say that the breast of veal, which is the part we had for lunch today, is a somewhat insipid dish when cooked English fashion. That we have been able to put it before you in more palatable form, and to win for it the approval of such a connoisseur as Sir John Oglethorpe, is largely owing to the judicious use of that Italian terror—more dire to many English than papermoney or brigands—garlic."

"The quantity used was infinitesimal," said Mrs. Sinclair, "but it seems to have been enough to subdue what I once heard Sir John describe as the pallid solidity of the innocent calf."

"I fear the vein of incongruity in our discourse, lately noted by Van der Roet, is not quite exhausted," said Sir John. "The Colonel was up in arms on account of a too intimate association of his name with pepper, and now Mrs. Sinclair has bracketed me with the calf, a most useful animal, I grant, but scarcely one I should have chosen as a yokefellow; but this is a digression. To return to our veal. I had a notion that garlic had something to do with the triumph of the Tenerumi, and, this being the case, I think it would be well if the Marchesa were to give us a dissertation on the use of this invaluable product."

"As Mrs. Sinclair says, the admixture of garlic in the dish in question was a very small one, and English people somehow never seem to realise that garlic must always be used sparingly. The chief positive idea they have of its characteristics is that which they gather from the odour of a French or Italian crowd of peasants at a railway station. The effect of garlic, eaten in lumps as an accompaniment to bread and cheese, is naturally awful, but garlic used as it should be used is the soul, the divine essence, of cookery. The palate delights in it without being able to identify it, and the surest proof of its charm is manifested by the flatness and insipidity which will infallibly characterise any dish usually flavoured with it, if by chance this dish should be prepared without it. The cook who can employ it successfully will be found to possess the delicacy of perception, the accuracy of judgment, and the dexterity of hand, which go to the formation of a great artist. It is a primary maxim, and one which cannot be repeated too often, that garlic must never be cut up and used as part of the material of any dish. One small incision should be made in the clove, which should be put into the dish during the process of cooking, and allowed to remain there until the cook's palate gives warning that flavour enough has been extracted. Then it must be taken out at once. This rule does not apply in equal degree to the use of the onion, the large mild varieties of which may be cooked and eaten in many excellent bourgeois dishes; but in all fine cooking, where the onion flavour is wanted, the same treatment which I have prescribed for garlic must be followed."

The Marchesa gave the Colonel and Lady Considine a holiday that afternoon, and requested Mrs. Gradinger and Van der Roet to attend in the kitchen to help with the dinner. In the first few days of the session the main portion of the work naturally fell upon the Marchesa and Angelina, and in spite of the inroads made upon their time by the necessary directions to the neophytes, and of the occasional eccentricities of the neophytes' energies, the dinners and luncheons were all that could be desired. The Colonel was not quite satisfied with the flavour of one particular soup, and Mrs. Gradinger was of opinion that one of the entrees, which she wanted to superintend herself, but which the Marchesa handed over to Mrs. Sinclair, had a great deal too much butter in its composition. Her conscience revolted at the action of consuming in one dish enough butter to solace the breakfast-table of an honest working man for two or three days; but the faintness of these criticisms seemed to prove that every one was well satisfied with the rendering of the menu of the day.

> Menu — Lunch Tenerumi di Vitello. Breast of veal. Piccione alla minute. Pigeons, braized with liver, &c. Curry
>
> Menu — Dinner Zuppa alla nazionale. Soup alla nazionale. Salmone alla Genovese. Salmon alla Genovese. Costolette alla Costanza. Mutton cutlets alla Costanza. Fritto misto alla Villeroy. Lamb's fry alla Villeroy. Lattughe al sugo. Stuffed Lettuce. Dindo arrosto alla Milanese. Roast turkey alla Milanese. Crema montata alle fragole. Strawberry cream. Tartufi alla Dino. Truffles alla Dino.

The Third Day

"I observe, dear Marchesa," said Mrs. Fothergill at breakfast on Thursday morning, "that we still follow the English fashion in our breakfast dishes. I have a notion that, in this particular especially, we gross English show our inferiority to the more spirituelles nations of the Continent, and I always feel a new being after the light meal of delicious coffee and crisp bread and delicate butter the first morning I awake in dear Paris."

"I wonder how it happens, then, that two goes of fish, a plateful of omelette, and a round and a half of toast and marmalade are necessary to repair the waste of tissue in dear England?" Van der Roet whispered to Miss Macdonnell.

"It must be the gross air of England or the gross nature of the—"

The rest of Miss Macdonnell's remark was lost, as the Marchesa cried out in answer to Mrs. Fothergill, "But why should we have anything but English breakfast dishes in England? The defects of English cookery are manifest enough, but breakfast fare is not amongst them. In these England stands supreme; there is nothing to compare with them, and they possess the crowning merit of being entirely compatible with English life. I cannot say whether it may be the effect of the crossing, or of the climate on this side, or that the air of England is charged with some subtle stimulating quality, given off in the rush and strain of strenuous national life, but the fact remains that as soon as I find myself across the Channel I want an English breakfast. It seems that I am more English than certain of the English themselves, and I am sorry that Mrs. Fothergill has been deprived of her French roll and butter. I will see that you have it tomorrow, Mrs. Fothergill, and to make the illusion complete, I will order it to be sent to your room."

"Oh no, Marchesa, that would be giving too much trouble, and I am sure you want all the help in the house to carry out the service as exquisitely as you do," said Mrs. Fothergill hurriedly, and blushing as well as her artistic complexion would allow.

"I fancy," said Mrs. Sinclair, "that foreigners are taking to English breakfasts as well as English clothes. I noticed when I was last in Milan that almost every German or Italian ate his two boiled eggs for breakfast, the sign whereby the Englishman used to be marked for a certainty."

"The German would probably call for boiled eggs when abroad on account of the impossibility of getting such things in his own country. No matter how often you send to the kitchen for properly boiled eggs in Germany, the result is always the same cold slush," said Mrs. Wilding; "and I regret to find that the same plague is creeping into the English hotels which are served by German waiters."

"That is quite true," said the Marchesa; "but in England we have no time to concern ourselves with mere boiled eggs, delicious as they are. The roll of delicacies is long enough, or even too long without them. When I am in England, I always lament that we have only seven days a week and one breakfast a day, and when I am in Italy I declare that the reason why the English have overrun the world is because they eat such mighty breakfasts. Considering how good the dishes are, I wonder the breakfasts are not mightier than they are."

"It always strikes me that our national barrenness of ideas appears as plainly in our breakfasts as anywhere," said Mrs. Gradinger. "There is a monotony about them which—"

"Monotony!" interrupted the Colonel. "Why, I could dish you up a fresh breakfast every day for a month. Your conservative tendencies must be very strong, Mrs. Gradinger, if they lead you to this conclusion."

"Conservative! On the contrary, I—that is, my husband—always votes for Progressive candidates at every election," said Mrs. Gradinger, dropping into her platform intonation, at the sound of which consternation arose in every breast. "I have, moreover, a theory that we might reform our diet radically, as well as all other institutions; but before I expound this, I should like to say a few words on the waste of wholesome food which goes on. For instance, I went for a walk in the woods yesterday afternoon, where I came upon a vast quantity of fungi which our ignorant middle classes would pronounce to be poisonous, but which I—in common with every child of the intelligent working-man educated in a board school where botany is properly taught—knew to be good for food."

"Excuse me one moment," said Sir John, "but do they really use board-school children as tests to see whether toadstools are poisonous or not?"

"I do not think anything I said justified such an inference," said Mrs. Gradinger in the same solemn drawl; "but I may remark that the children are taught from illustrated manuals accurately drawn and coloured. Well, to come back to the fungi, I took the trouble to measure the plot on which they were growing, and found it just ten yards square. The average weight of edible fungus per square yard was just an ounce, or a hundred and twelve pounds per acre. Now, there must be at least twenty millions of acres in the United Kingdom capable of producing these fungi without causing the smallest damage to any other crop, wherefore it seems that, owing to our lack of instruction, we are wasting some million tons of good food per annum; and I may remark that this calculation pre-supposes, that each fungus springs only once in the season; but I have reason to believe that certain varieties would give five or six gatherings between May and October, so the weight produced would be enormously greater than the quantity I have named."

Here Mrs. Gradinger paused to finish her coffee, which was getting cold, and before she could resume, Sir John had taken up the parole. "I think the smaller weight will suffice for the present, until the taste for strange fungi has developed, or the pressure of population increased. And before stimulating a vastly increased supply, it will be necessary to extirpate the belief that all fungi, except the familiar mushroom, are poisonous, and perhaps to appoint an army of inspectors to see that only the right sort are brought to market."

"Yes, and that will give pleasant and congenial employment to those youths of the working-classes who are ambitious of a higher career than that of their fathers," said Lady Considine, "and the ratepayers will rejoice, no doubt, that they are participating in the general elevation of the masses."

"Perhaps Mrs. Gradinger will gather a few of her less deadly fungi, and cook them and eat them herself, pour encourager les autres," said Miss Macdonnell. "Then, if she doesn't die in agonies, we may all forswear beef and live on toadstools."

"I certainly will," said Mrs. Gradinger; "and before we rise from table I should like—"

"I fear we must hear your remarks at dinner, Mrs. Gradinger," said the Marchesa. "Time is getting on, and some of the dishes today are rather elaborate, so now to the kitchen."

> Menu—Lunch. Risotto alla Genovese. Savoury rice. Pollo alla Villereccia. Chicken alla Villereccia. Lingue di Castrato alla cucinira. Sheeps' tongues alla cucinira. Menu—Dinner Zuppa alla Veneziana. Venetian soup. Sogliole alla giardiniera. Sole with Vegetables. Timballo alla Romana. Roman pie. Petto di Castrato alla salsa di burro. Breast of mutton with butter sauce. Verdure miste. Mixed vegetables. Crema rappresa. Coffee cream. Ostriche alla Veneziana. Oyster savoury.

The Fourth Day

THE Colonel was certainly the most severely critical member of the company. Up to the present juncture he had been sparing of censure, and sparing of praise likewise, but on this day, after lunch, he broke forth into loud praise of the dish of beef which appeared in the menu. After specially commending this dish he went on—

"It seems to me that the dinner of yesterday and to-day's lunch bear the cachet of a fresh and admirable school of cookery. In saying this I don't wish to disparage the traditions which have governed the preparation of the delicious dishes put before us up to that date, which I have referred to as the parting of the ways, the date when the palate of the expert might detect a new hand upon the keys, a phrase once employed, I believe, with regard to some man who wrote poetry. To meet an old friend, or a thoroughly tested dish, is always pleasant, but old friends die or fall out, and old favourite dishes may come to pall at last; and for this reason I hold that the day which brings us a new friend or a new dish ought to be marked with white chalk."

"And I think some wise man once remarked," said Sir John, "that the discovery of a dish is vastly more important than the discovery of a star, for we have already as many stars as we can possibly require, but we can never have too many dishes."

"I was wondering whether any one would detect the variations I made yesterday, but I need not have wondered, with such an expert at table as Colonel Trestrail," said the Marchesa with a laugh. "Well, the Colonel has found me out; but from the tone of his remarks I think I may hope for his approval. At any rate, I'm sure he won't move a vote of censure."

"If he does, we'll pack him off to town, and sentence him to dine at his club every day for a month," said Lady Considine.

"What crime has this particular club committed?" said Mrs. Sinclair in a whisper.

"Vote of censure! Certainly not," said the Colonel, with an angry ring in his voice. Mrs. Sinclair did not love him, and had calculated accurately the carrying power of her whisper. "That would be the basest ingratitude. I must, however, plead guilty to an attack of curiosity, and therefore I beg you, Marchesa, to let us into the secret of your latest inspiration."

"Its origin was commonplace enough," said the Marchesa, "but in a way interesting. Once upon a time—more years ago than I care to remember—I was strolling about the Piazza Navona in Rome, and amusing myself by going from one barrow to another, and turning over the heaps of rubbish with which they were stocked. All the while I was innocently plagiarising that fateful walk of Browning's round the Riccardi Palace in Florence, the day when he bought for a lira the Romana homocidiorum. The world knows what was the outcome of Browning's purchase, but it will probably never fathom the full effect of mine. How do his lines run?"

> "These I picked the book from. Five compeers in flank Stood left and right of it as tempting more— A dog's-eared Spicilegium, the fond tale O' the frail one of the Flower, by young Dumas, Vulgarised Horace for the use of schools, The Life, Death, Miracles of Saint Somebody, Saint Somebody Else, his Miracles, Death and Life."

"Well, the choice which lay before me on one particular barrow was fully as wide, or perhaps wider than that which met the poet's eye, but after I had espied a little yellow paper-covered book with the title La Cucina Partenopea, overo il Paradiso dei gastronomi, I looked no farther. What infinite possibilities of pleasure might lie hidden under such a name. I secured it, together with the Story of Barlaam and Josaphat, for thirty-five centesimi, and handed over the coins to the hungry-eyed old man in charge, who regretted, I am sure, when he saw the eager look upon my face, that he had not marked the books a lira at least. I should now be a rich woman if I had spent all the money I have spent as profitably as those seven sold. Besides being a master in the art of cookery, the author was a moral philosopher as well; and he addresses his reader in prefatory words which bespeak a profound knowledge of life. He writes:

'Though the time of man here on earth is passed in a never-ending turmoil, which must make him often curse the moment when he opened his eyes on such a world; though life itself must often become irksome or even intolerable, nevertheless, by God's blessing, one supreme consolation remains for this wretched body of ours. I allude to that moment when, the forces being spent and the stomach craving support, the wearied mortal sits down to face a good dinner. Here is to be found an effectual balm for the ills of life: something to drown all remembrance of our ill-humours, the worries of business, or even family quarrels. In sooth, it is only at table that a man may bid the devil fly away with Solomon and all his wisdom, and give himself up to an earthly delight, which is a pleasure and a profit at the same time.'"

"The circumstances under which this precious book was found seem to suggest a culinary poem on the model of the 'Ring and the Book,'" said Mrs. Sinclair, "or we might deal with the story in practical shape by letting every one of us prepare the same dish. I fancy the individual renderings of the same recipe would vary quite as widely as the versions of the unsavoury story set forth in Mr. Browning's little poem."

"I think we had better have a supplementary day for a trial of the sort Mrs. Sinclair suggests," said Miss Macdonnell. "I speak with the memory of a preparation of liver I tasted yesterday in the kitchen — one of the dishes which did not appear at dinner."

"That is rather hard on the Colonel," said Van der Roet; "he did his best, and now, see how hard he is trying to look as if he didn't know what you are alluding to!"

"I never in all my life—" the Colonel began; but the Marchesa, fearing a storm, interfered. "I have a lot more to tell you about my little Neapolitan book," she went on, "and I will begin by saying that, for the future, we cannot do better than make free use of it. The author opens with an announcement that he means to give exact quantities for every dish, and then, like a true Neapolitan, lets quantities go entirely, and adopts the rule-of-thumb system. And I must say I always find the question of quantities a difficult one. Some books give exact measures, each dish being reckoned enough for four persons, with instructions to increase the measures in proportion to the additional number of diners but here a rigid rule is impossible, for a dish which is to serve by itself, as a supper or a lunch, must necessarily be bigger than one which merely fills one place in a dinner menu. Quantities can be given approximately in many cases, but flavouring must always be a question of individual taste. Latitude must be allowed, for all cooks who can turn out distinguished work will be found to be endowed with imagination, and these, being artists, will never consent to follow a rigid rule of quantity. To put it briefly, cooks who need to be told everything, will never cook properly, even if they be told more than everything. And after all, no one takes seriously the quantities given by the chef of a millionaire or a prince; witness the cook of the Prince de Soubise, who demanded fifty hams for the sauces and garnitures of a single supper, and when the Prince protested that there could not possibly be found space for them all on the table, offered to put them all into a glass bottle no bigger than his thumb. Some of Francatelli's quantities are also prodigious, as, for instance, when to make a simple glaze he calls for three pounds of gravy beef, the best part of a ham, a knuckle of veal, an old hen, and two partridges."

> Menu — Lunch Maccheroni al sugillo. Macaroni with sausage and tomatoes. Manzo in insalata. Beef, pressed and marinated. Lingue di vitello all'Italiana. Calves' tongues.

> Menu — Dinner. Zuppa alla Modanese. Modenese soup. Merluzzo in salamoia. Cod with sauce piquante. Pollastro in istufa di pomidoro. Stewed chicken with tomatoes. Porcelletto farcito alla Corradino. Stuffed suckling pig. Insalata alla Navarino. Navarino salad. Bodino di semolino. Semolina pudding. Frittura di cocozze. Fried cucumber.

The Fifth Day

The following day was very warm, and some half-dozen of the party wandered into the garden after lunch and took their coffee under a big chestnut tree on the lawn. "And this is the 16th of June," said Lady Considine. "Last year, on this very day, I started for Hombourg. I can't say I feel like starting for Hombourg, or any other place, just at present."

"But why should any one of us want to go to Hombourg?" said Sir John. "Nobody can be afraid of gout with the admirable diet we enjoy here."

"I beg you to speak for yourself, Sir John," said Lady Considine. "I have never yet gone to Hombourg on account of gout."

"Of course not, my dear friend, of course not; there are so many reasons for going to Hombourg. There's the early rising, and the band, and the new people one may meet there, and the change of diet—especially the change of diet. But, you see, we have found our change of diet within an hour of London, so why—as I before remarked—should we want to rush off to Hombourg?"

"I am a firm believer in that change of diet," said Mrs. Wilding, "though in the most respectable circles the true-bred Briton still talks about foreign messes, and affirms that anything else than plain British fare ruins the digestion. I must say my own digestion is none the worse for the holiday I am having from the preparations of my own 'treasure.' I think we all look remarkably well; and we don't quarrel or snap at each other, and it would be hard to find a better proof of wholesome diet than that."

"But I fancied Mrs. Gradinger looked a little out of sorts this morning, and I'm sure she was more than a little out of temper when I asked her how soon we were to taste her dish of toadstools," said Miss Macdonnell.

"I expect she had been making a trial of the British fungi in her bedroom," said Van der Roet; "and then, you see, our conversation isn't quite 'high toned' enough for her taste. We aren't sufficiently awake to the claims of the masses. Can any one explain to me why the people who are so full of mercy for the mass, are so merciless to the unit?"

"That is her system of proselytising," said the Colonel, "and if she is content with outward conversion, it isn't a bad one. I often feel inclined to agree to any proposition she likes to put forward, and I would, if I could stop her talking by my submission."

"You wouldn't do that, Colonel, even in your suavest mood," said Van der Roet; "but I hope somebody will succeed in checking her flow of discourse before long. I'm getting worn to a shadow by the grind of that awful voice."

"I thought your clothes were getting a bit loose," said the Colonel, "but I put that phenomenon down to another reason. In spite of Mrs. Wilding's praise of our present style of cooking, I don't believe our friend Vander finds it substantial enough to sustain his manly bulk, and I'll tell you the grounds of my belief. A few mornings ago, when I was shaving, I saw the butcher bring into the house a splendid sirloin, and as no sirloin has appeared at table, I venture to infer that this joint was a private affair of Vander's, and that he, as well as Mrs. Gradinger, has been going in for bedroom cookery. Here comes the Marchesa; we'll ask her to solve the mystery."

"I can account for the missing sirloin," said the Marchesa. "The Colonel is wrong for once. It went duly into the kitchen, and not to Mr. Van der Roet's bedroom; but I must begin with a slight explanation, or rather apology. Next to trial by jury, and the reverence paid to rank, and the horror of all things which, as poor Corney Grain used to say, 'are not nice,' I reckon the Sunday sirloin, cooked and served, one and indivisible as the typical fetish of the great English middle class. With this fact before my eyes, I can assure you I did not lightly lay a hand on its integrity. My friends, you have eaten that sirloin without knowing it. You may remember that yesterday after lunch the Colonel was loud in praise of a dish of beef. Well, that beef was a portion of the same, and not the best portion. The Manzo in insalata, which pleased the Colonel's palate, was that thin piece at the lower end, the chief function of which, when the sirloin is cooked whole, seems to lie in keeping the joint steady on the dish while paterfamilias carves it. It is never eaten in the dining-room hot, because every one justly prefers and goes for the under cut; neither does it find favour at lunch next day, for the reason that, as cold beef, the upper cut is unapproachable. I have never heard that the kitchen hankers after it inordinately; indeed, its ultimate destination is one of the unexplained mysteries of housekeeping. I hold that never, under any circumstances, should it be cooked with the sirloin, but always cut off and marinated and braized as we had it yesterday. Thus you get two hot dishes; our particular sirloin has given us three. The parts of this joint vary greatly in flavour, and in texture as well, and by accentuating this variation by treatment in the kitchen, you escape that monotony which is prone to pervade the table so long as the sirloin remains in the house. Mrs. Sinclair is sufficiently experienced as a housekeeper to know that the dish of fillets we had for dinner last night was not made from the under cut of one sirloin. It was by borrowing a little from the upper part that I managed to fill the dish, and I'm sure that any one who may have got one of the uppercut fillets had no cause to grumble. The Filetto di Bue which we had for lunch to-day was the residue of the upper cut, and, admirable as is a slice of cold beef taken from this part of the joint, I think it is an excellent variation to make a hot dish of it sometimes. On the score of economy, I am sure that a sirloin treated in this fashion goes a long way further."

"The Marchesa demolishes one after another of our venerable institutions with so charming a despatch that we can scarcely grieve for them," said Sir John. "I am not philosopher enough to divine what change may come over the British character when every man sits down every day to a perfectly cooked dinner. It is sometimes said that our barbarian forefathers left their northern solitudes because they hankered after the wine and delicate meats of the south, and perhaps the modern Briton may have been led to overrun the world by the hope of finding a greater variety of diet than he gets at home. It may mean, Marchesa, that this movement of yours for the suppression of English plain cooking will mark the close of our national expansion."

"My dear Sir John, you may rest assured that your national expansion, as well as your national cookery, will continue in spite of anything we may accomplish here, and I say good luck to them both. When have I ever denied the merits of English cookery?" said the Marchesa. "Many of its dishes are unsurpassed. These islands produce materials so fine, that no art or elaboration can improve them. They are best when they are cooked quite plainly, and this is the reason why simplicity is the key-note of English cookery. A fine joint of mutton roasted to a turn, a plain fried sole with anchovy butter a broiled chop or steak or kidney, fowls or game cooked English fashion, potatoes baked in their skins and eaten with butter and salt, a rasher of Wiltshire bacon and a new-laid egg, where will you beat these? I will go so far as to say no country can produce a bourgeoises dish which can be compared with steak and kidney pudding. But the point I want to press home is that Italian cookery comes to the aid of those who cannot well afford to buy those prime qualities of meat and fish which allow of this perfectly plain treatment. It is, as I have already said, the cookery of a nation short of cash and unblessed with such excellent meat and fish and vegetables as you lucky islanders enjoy. But it is rich in clever devices of flavouring, and in combinations, and I am sure that by its help English people of moderate means may fare better and spend less than they spend now, if only they will take a little trouble."

> Menu — Lunch Gnocchi alla Romana. Semolina with parmesan. Filetto di Bue al pistacchi. Fillet of beef with pistachios Bodini marinati. Marinated rissoles.
>
> Menu — Dinner. Zuppa Crotopo. Croute au pot soup. Sogliole alla Veneziana. Fillets of sole. Ateletti alla Sarda. Atelets of ox-palates, &c. Costolette di Montone alla Nizzarda. Mutton cutlets. Pollo alla Fiorentina. Fowl with macaroni. Crema tartara alla Caramella. Caramel cream. Uova rimescolati al tartufi. Eggs with truffles.

The Sixth Day

The following morning, at breakfast, a servant announced that Sir John Oglethorpe was taking his breakfast in his room, and that there was no need to keep anything in reserve for him. It was stated, however, that Sir John was in no way indisposed, and that he would join the party at lunch.

He seated himself in his usual place, placid and fresh as ever; but, unharmed as he was physically, it was evident to all the company that he was suffering from some mental discomposure. Miss Macdonnell, with a frank curiosity which might have been trying in any one else, asked him point-blank the reason of his absence from the meal for which, in spite of his partiality for French cookery, he had a true Englishman's devotion.

"I feel I owe the company some apology for my apparent churlishness," he said; "but the fact is, that I have received some very harrowing, but at the same time very interesting, news this morning. I think I told you the other day how the vacancy in my kitchen has led up to a very real tragedy, and that the abhorred Fury was already hovering terribly near the head of poor Narcisse. Well, I have just received from a friend in Paris journals containing a full account of the trial of Narcisse and of his fair accomplice. The worst has come to pass, and Narcisse has been doomed to sneeze into the basket like a mere aristocrat or politician during the Terror I was greatly upset by this news, but I was interested, and in a measure consoled, to find an enclosure amongst the other papers, an envelope addressed to me in the handwriting of the condemned man. This voix d'outre tombe, I rejoice to say, confides to me the secret of that incomparable sauce of his, a secret which I feared might be buried with Narcisse in the prison ditch."

The Marchesa sighed as she listened. The recipe of the sauce was safe indeed, but she knew by experience how wide might be the gulf between the actual work of an artist and the product of another hand guided by his counsels, let the hand be ever so dexterous, and the counsels ever so clear. "Will it be too much," she said, "to ask you to give us the details of this painful tragedy?"

"It will not," Sir John replied reflectively. "The last words of many a so-called genius have been enshrined in literature: probably no one will ever know the parting objurgation of Narcisse. I will endeavour, however, to give you some notion as to what occurred, from the budget I have just read. I fear the tragedy was a squalid one. Madame, the victim, was elderly, unattractive in person, exacting in temper, and the owner of considerable wealth—at least, this is what came out at the trial. It was one of those tangles in which a fatal denouement is inevitable; and, if this had not come through Mademoiselle Sidonie, it would have come through somebody else.

The lovers plotted to remove madame by first drugging her, then breaking her skull with the wood chopper, and then pitching her downstairs so as to produce the impression that she had met her death in this fashion. But either the arm of Mademoiselle Sidonie—who was told off to do the hammering—was unskilled in such work, or the opiate was too weak, for the victim began to shriek before she gave up the ghost. Detection seemed imminent, so Narcisse, in whom the quality of discretion was evidently predominant, bolted at once and got out of the country. But the facts were absolutely clear. The victim lived long enough to depose that Mademoiselle Sidonie attacked her with the wood chopper, while Narcisse watched the door. The advocate of Narcisse did his work like a man. He shed the regulation measure of tears; he drew graphic pictures of the innocent youth of Narcisse, of his rise to eminence, and of his filial piety as evidenced by the frequent despatch of money and comestibles to his venerable mother, who was still living near Bourges. Once a year, too, this incomparable artist found time to renew his youth by a sojourn in the simple cottage which saw his birth, and by embracing the giver of his life. Was it possible that a man who treated one woman with such devotion and reverence could take the life of another? He adduced various and picturesque reasons to show that such an event must be impossible, but the jury took the opposite view. Some one had to be guillotined, and the intelligent jury decided that Paris could spare Narcisse better than it could spare Mademoiselle Sidonie. I fear the fact that he had deigned to sell his services to a brutal islander may have helped them to come to this conclusion, but there were other and more weighty reasons. Of the supreme excellence of Narcisse as an artist the jury knew nothing, so they let him go hang—or worse—but of Mademoiselle Sidonie they knew a good deal, and their knowledge, I believe, is shared by certain English visitors to Paris. She is one of the attractions of the Fantasies d'Arcadie, and her latest song, Bonjour Coco, is sung and whistled in every capital of Europe; so the jury, thrusting aside as mere pedantry the evidence of facts, set to work to find some verdict which would not eclipse the gaiety of La Ville Lumiere by cutting short the career of Mademoiselle Sidonie. The art of the chef appealed to only a few, and he dies a mute, but by no means inglorious martyr: the art of the chanteuse appeals to the million, the voice of the many carries the day, and Narcisse must die."

"It is a revolting story," said Mrs. Gradinger, "and one possible only in a corrupted and corrupting society. It is wonderful, as Sir John remarks, how the conquering streams of tendency manifest themselves even in an affair like this. Ours is a democratic age, and the wants and desires of the many, who find delight in this woman's singing, override the whims of the pampered few, the employers of such costly luxuries as men cooks."

"You see you are a mere worm, Sir John," laughed Miss Macdonnell, "and you had better lay out your length to be trampled on."

"Yes, I have long foreseen our fate, we who happen to possess what our poor brother hankers after. Well, perhaps I may take up the worm's role at once and 'turn', that is, burn the recipe of Narcisse."

"O Sir John, Sir John," cried Mrs. Sinclair "any such burning would remind me irresistibly of Mr. Mantalini's attempts at suicide. There would be an accurate copy in your pocket-book, and besides this you would probably have learnt off the recipe by heart."

"Yes, we know our Sir John better than that, don't we?" said the Marchesa; "but, joking apart, Sir John, you might let me have the recipe at once. It would go admirably with one of our lunch dishes for to-morrow."

But on the subject of the sauce, Sir John—like the younger Mr. Smallweed on the subject of gravy—was adamant. The wound caused by the loss of Narcisse was, he declared, yet too recent: the very odour of the sauce would provoke a thousand agonising regrets. And then the hideous injustice of it all: Narcisse the artist, comparatively innocent (for to artists a certain latitude must be allowed), to moulder in quicklime, and this greedy, sordid murderess to go on ogling and posturing with superadded popularity before an idiot crowd unable to distinguish a Remoulade from a Ravigotte! "No, my dear Marchesa," he said, "the secret of Narcisse must be kept a little longer, for, to tell the truth, I have an idea. I remember that ere this fortunes have been made out of sauces, and if this sauce be properly handled and put before the public, it may counteract my falling, or rather disappearing rents. If only I could hit upon a fetching name, and find twenty thousand pounds to spend in advertising, I might be able once more to live on my acres."

"Oh, surely we shall be able to find you a name between us," said Mrs. Wilding; "money, and things of that sort are to be procured in the city, I believe; and I daresay Mr. Van der Roet will design a pretty label for the sauce bottles."

> Menu—Lunch. Pollo all'olive. Fowl with olives. Scaloppine di rive. Veal cutlets with rice. Sedani alla parmigiana. Stewed celery.
>
> Menu—Dinner. Zuppa primaverile. Spring soup Sote di Salmone al funghi. Salmon with mushrooms. Tenerumi d'Agnello alla veneziana. Breast of lamb alla Veneziana. Testa di Vitello alla sorrentina. Calf's head alla Sorrentina. Fagiano alla perigo. Pheasant with truffles. Torta alla cremonese. Cremona tart. Uova alla fiorentina. Egg savoury.

The Seventh Day

"It seems invidious to give special praise where everything is so good," said Mrs. Sinclair next day at lunch, "but I must say a word about that clear soup we had at dinner last night. I have never ceased to regret that my regard for manners forbade me ask for a second helping."

"See what it is to have no manners," said Van der Roet. "I plunged boldly for another portion of that admirable preparation of calf's head at dinner. If I hadn't, I should have regretted it for ever after. Now, I'm sure you are just as curious about the construction of these masterpieces as I am, Mrs. Sinclair, so we'll beg the Marchesa to let us into the secret."

"Mrs. Sinclair herself had a hand in the calf's-head dish, 'Testa di Vitello alla sorrentina,' so perhaps I may hand over that part of the question to her. I am very proud that one of my pupils should have won praise from such a distinguished expert as Mr. Van der Roet, and I leave her to expound the mystery of its charm. I think I may without presumption claim the clear soup as a triumph, and it is a discovery of my own. The same calf's head which Mrs. Sinclair has treated with such consummate skill, served also as the foundation for the stock of the clear soup. This stock certainly derived its distinction from the addition of the liquor in which the head was boiled. A good consomme can no doubt be made with stock-meat alone, but the best soup thus made will be inferior to that we had for dinner last night. Without the calf's head you will never get such softness, combined with full roundness on the tongue, and the great merit of calf's head is that it lets you attain this excellence without any sacrifice of transparency."

"I have marvelled often at the clearness of your soups, Marchesa," said the Colonel. "What clearing do you use to make them look like pale sherry?"

"No one has any claim to be called a cook who cannot make soup without artificial clearing," said the Marchesa. "Like the poet, the consomme is born, not made. It must be clear from the beginning, an achievement which needs care and trouble like every other artistic effort, but one nevertheless well within the reach of any student who means to succeed. To clear a soup by the ordinary medium of white of egg or minced beef is to destroy all flavour and individuality. If the stock be kept from boiling until it has been strained, it will develop into a perfectly clear soup under the hands of a careful and intelligent cook. The fleeting delicate aroma which, as every gourmet will admit, gives such grateful aid to the palate, is the breath of garden herbs and of herbs alone, and here I have a charge to bring against contemporary cookery. I mean the neglect of natural in favour of manufactured flavourings. With regard to herbs, this could not always have been the rule, for I never go into an old English garden without finding there a border with all the good old-fashioned pot herbs growing lustily. I do not say that the use of herbs is unknown, for of course the best cookery is impossible without them, but I fear that sage mixed with onion is about the only one which ever tickles the palate of the great English middle-class. And simultaneously with the use of herb flavouring in soup has arisen the practice of adding wine, which to me seems a very questionable one. If wine is put in soup at all, it must be used so sparingly as to render its presence imperceptible. Why then use it at all? In some sauces wine is necessary, but in all cases it is as difficult to regulate as garlic, and requires the utmost vigilance on the part of the cook."

"My last cook, who was very stout and a little middle-aged, would always use flavouring sauces from the grocer's rather than walk up to the garden, where we have a most seductive herb bed," said Mrs. Wilding; "and then, again, the love of the English for pungent-made sauces is another reason for this makeshift practice. 'Oh, a table-spoonful of somebody's sauce will do for the flavouring,' and in goes the sauce, and the flavouring is supposed to be complete. People who eat their chops, and steaks, and fish, and game, after having smothered the natural flavour with the same harsh condiment, may be satisfied with a cuisine of this sort, but to an unvitiated palate the result is nauseous."

"Yet as a Churchwoman, Mrs. Wilding, you ought to speak with respect of English sauces. I think I have heard how a libation of one of them, which was poured over a certain cathedral, has made it look as good as new," said Miss Macdonnell, "and we have lately learned that one of the most distinguished of our party is ambitious to enter the same career."

"I would suggest that Sir John should devote all that money he proposes to make by the aid of his familiar spirit—the ghost of Narcisse—to the building of a temple in honour of the tenth muse, the muse of cookery," said Mrs. Sinclair; "and what do you think, Sir John, of a name I dreamt of last night for your sauce, 'The New Century Sauce'? How will that do?"

"Admirably," said Sir John after a moment's pause; "admirably enough to allow me to offer you a royalty on every bottle sold. 'The New Century Sauce', that's the name for me; and now to set to work to build the factory, and to order plans for the temple of the tenth muse."

> Menu—Lunch. Maccheroni al pomidoro. Macaroni with tomatoes, Vitello alla pellegrina. Veal cutlets alla pellegrina.

Animelle al sapor di targone. Sweetbread with tarragon sauce.

Menu — Dinner. Zuppa alla Canavese. Soup alla Canavese Naselli con piselli. Whiting with peas. Coscia di manzo al forno. Braized ribs of beef. Lingua alla Visconti. Tongue with grapes. Anitra selvatica. Wild duck. Zabajone ghiacciato. Iced syllabub. Crostatini alla capucina. Savoury of rice, truffles, &c.

The Eighth Day

"We are getting unpleasantly near the end of our time," said the Colonel, "but I am sure not one of us has learnt one tithe of what the Marchesa has to teach."

"My dear Colonel Trestrail," said the Marchesa, "an education in cookery does not mean the teaching of a certain number of recipes. Education, I maintain, is something far higher than the mere imparting of facts; my notion of it is the teaching of people to teach themselves, and this is what I have tried to do in the kitchen. With some of you I am sure I have succeeded, and a book containing the recipe of every dish we have tried will be given to every pupil when we break up."

"I think the most valuable lesson I have learnt is that cookery is a matter for serious study," said Mrs. Sinclair. "The popular English view seems to be that it is one of those things which gets itself done. The food is subjected to the action of heat, a little butter, or pepper, or onion, being added by way of flavouring, and the process is complete. To put it bluntly, it requires at least as much mental application to roast a fowl as to cut a bodice; but it does not strike the average Englishwoman in this way, for she will spend hours in thinking and talking about dressmaking (which is generally as ill done as her cooking), while she will be reluctant to give ten minutes to the consideration as to how a luncheon or supper dish shall be prepared. The English middle classes are most culpably negligent about the food they eat, and as a consequence they get exactly the sort of cooks they deserve to get. I do not blame the cooks; if they can get paid for cooking ill, why should they trouble to learn to cook well?"

"I agree entirely," said Mrs. Wilding. "That saying, 'What I like is good plain roast and boiled, and none of your foreign kickshaws,' is, as every one knows, the stock utterance of John Bull on the stage or in the novel; and, though John Bull is not in the least like his fictitious presentment, this form of words is largely responsible for the waste and want of variety in the English kitchen. The plain roast and boiled means a joint every day, and this arrangement the good plain cook finds an admirable one for several reasons: it means little trouble, and it means also lots of scraps and bones and waste pieces. The good plain cook brings all the forces of obstruction to bear whenever the mistress suggests made dishes; and, should this suggestion ever be carried out, she takes care that the achievement shall be of a character not likely to invite repetition. Not long ago a friend of mine was questioning a cook as to soups, whereupon the cook answered that she had never been required to make such things where she had lived; all soups were bought in tins or bottles, and had simply to be warmed up. Cakes, too, were outside her repertoire, having always been 'had in' from the confectioner's, while 'entrys' were in her opinion, and in the opinion of her various mistresses, 'un'ealthy' and not worth making."

"My experience is that, if a mistress takes an interest in cooking, she will generally have a fairly efficient cook," said Mrs. Fothergill. "I agree with Mrs. Sinclair that our English cooks are spoilt by neglect; and I think it is hard upon them, as a class, that so many inefficient women should be able to pose as cooks while they are unable to boil a potato properly."

"And the so-called schools of cookery are quite useless in what they teach," said Miss Macdonnell. "I once sent a cook of mine to one to learn how to make a clear soup, and when she came back, she sent up, as an evidence of her progress, a potato pie coloured pink and green, a most poisonous-looking dish—and her clear soups were as bad as ever."

Said the Colonel, "I will beg leave to enter a protest against the imperfections of that repast which is supposed to be the peculiar delight of the ladies, I allude to afternoon tea. I want to know why it is that unless I happen to call just when the tea is brought up—I grant, I know of a few houses which are honourable exceptions—I am fated to drink that most abominable of all decoctions, stewed lukewarm tea. 'Will you have some tea? I'm afraid it isn't quite fresh,' the hostess will remark without a blush. What would she think if her husband at dinner were to say, 'Colonel, take a glass of that champagne. It was opened the day before yesterday, and I daresay the fizz has gone off a little'? Tea is cheap enough, and yet the hostess seldom or never thinks of ordering up a fresh pot. I believe it is because she is afraid of the butler."

"I sympathise with you fully, Colonel," said Lady Considine, "and my withers are unwrung. You do not often honour me with your presence on Tuesdays, but I am sure I may claim to be one of your honourable exceptions."

"Indeed you may," said the Colonel. "Perhaps men ought not to intrude on these occasions; but I have a preference for taking tea in a pretty drawing-room, with a lot of agreeable women, rather than in a club surrounded by old chaps growling over the latest job at the War Office, and a younger brigade chattering about the latest tape prices, and the weights for the spring handicaps."

"All these little imperfections go to prove that we are not a nation of cooks," said Van der Roet. "We can't be everything. Heine once said that the Romans would never have found time to conquer the world if they had been obliged to learn the Latin grammar; and it is the same with us. We can't expect to found an empire all over the planet, and cook as well as the French, who—perhaps wisely—never willingly emerge from the four corners of their own land."

"There is energy enough left in us when we set about some purely utilitarian task," said Mrs. Wilding, "but we never throw ourselves into the arts with the enthusiasm of the Latin races. I was reading the other day of a French costumier who rushed to inform a lady, who had ordered a turban, of his success, exclaiming, 'Madame, apres trots nun's d'insomnie les plumes vent placees.' And every one knows the story of Vatel's suicide because the fish failed to arrive. No Englishman would be capable of flights like these."

"Really, this indictment of English cookery makes me a little nervous," said Lady Considine "I have promised to join in a driving tour through the southern counties. I shudder to think of the dinners I shall have to eat at the commercial hotels and posting-houses on our route."

"English country inns are not what they ought to be, but now and then you come across one which is very good indeed, as good, if not better, than anything you could find in any other country; but I fear I must admit that, charges considered, the balance is against us," said Sir John.

"When you start you ought to secure Sir John's services as courier, Lady Considine," said the Marchesa. "I once had the pleasure of driving for a week through the Apennines in a party under his guidance, and I can assure you we found him quite honest and obliging."

"Ah, Marchesa, I was thinking of that happy time this very morning," said Sir John. "Of Arezzo, where we were kept for three days by rain, which I believe is falling there still. Of Cortona, with that wonderful little restaurant on the edge of the cliff, whence you see Thrasumene lying like a silver mirror in the plain below. Of Perugia, the august, of Gubbio, Citta di Castello, Borgo San Sepolcro, Urbino, and divers others. If you go for a drive in Italy, you still may meet with humours of the road such as travellers of old were wont to enjoy. I well remember on the road between Perugia and Gubbio we began to realise we were indeed traversing mountain paths. On a sudden the driver got down, waved his arms, and howled to some peasants working in a field below. These, on their part, responded with more arm-waving and howling, directed apparently towards a village farther up the hill, whereupon we were assailed with visions of brigands, and amputated ears, and ransom. But at a turn of the road we came upon two magnificent white oxen, which, being harnessed on in front, drew us, and our carriages and horses as well, up five miles of steep incline. These beautiful fellows, it seemed, were what the driver was signalling for, and not for brigands. Again, every inn we stayed at supplied us with some representative touch of local life and habit. Here the whole personnel of the inn, reinforced by a goodly contingent of the townsfolk, would accompany us even into our bedrooms, and display the keenest interest in the unpacking of our luggage. There the cook would come and take personal instructions as to the coming meal, throwing out suggestions the while as to the merits of this or that particular dish, and in one place the ancient chambermaid insisted that one of the ladies, who had got a slight cold, should have the prete put into her bed for a short time to warm it. You need not look shocked, Colonel. The prete in question was merely a wooden frame, in the midst of which hangs a scaldino filled with burning ashes — a most comforting ecclesiastic, I can assure you. All the inns we visited had certain characteristics in common. The entrance is always dirty, and the staircase too, the dining rooms fairly comfortable, the bedrooms always clean and good, and the food much better than you would expect to find in such out-of-the-way places; indeed I cannot think of any inn where it was not good and wholesome, while often it was delicious. In short, Lady Considine, I strongly advise you to take a drive in Italy next spring, and if I am free I shall be delighted to act as courier."

"Sir John has forgotten one or two touches I must fill in," said the Marchesa. "It was often difficult to arrange a stopping-place for lunch, so we always stocked our basket before starting. After the first day's experience we decided that it was vastly more pleasant to take our meal while going uphill at a foot-pace, than in the swing and jolt of a descent, so the route and the pace of the horses had to be regulated in order to give us a good hour's ascent about noon. Fortunately hills are plentiful in this part of Italy, and in the keen air we generally made an end of the vast store of provisions we laid in, and the generous fiasco was always empty a little too soon. Our drive came to an end at Fano, whither we had gone on account of a strange romantic desire of Sir John to look upon an angel which Browning had named in one of his poems. Ah! how vividly I can recall our pursuit of that picture. It was a wet, melancholy day. The people of Fano were careless of the fame of their angel, for no one knew the church which it graced. At last we came upon it by the merest chance, and Sir John led the procession up to the shrine, where we all stood for a time in positions of mock admiration. Sir

John tried hard to keep up the imposition, but something, either his innate honesty or the chilling environment of disapproval of Guercino's handiwork, was too much for him. He did his best to admire, but the task was beyond his powers, and he raised no protest when some scoffer affirmed that, though Browning might be a great poet, he was a mighty poor judge of painting, when he gave in his beautiful poem immortality to this tawdry theatrical canvas. 'I think,' said Sir John, 'we had better go back to the hotel and order lunch. It would have been wiser to have ordered it before we left.' We were all so much touched by his penitence that no one had the heart to remind him how a proposition as to lunch had been made by our leading Philistine as soon as we arrived, a proposition waved aside by Sir John as inadmissible until the 'Guardian Angel' should have been seen and admired."

"I plead guilty," said Sir John. "I think this experience gave a death-blow to my career as an appreciator. Anyhow, I quite forget what the angel was like, and for reminiscences of Fano have to fall back upon the excellent colazione we ate in the externally unattractive, but internally admirable, Albergo del Moro."

> Menu—Lunch. Astachi all'Italiana. Lobster all'Italiana Filetto di bue alla Napolitana. Fillet of beef with Neapolitan sauce. Risotto alla spagnuola. Savoury rice.
>
> Menu—Dinner. Zuppa alla Romana. Soup with quenelles. Salmone alla Genovese. Salmon alla Genovese. Costolette in agro-dolce. Mutton cutlets with Roman sauce. Flano di spinacci. Spinach in a mould. Cappone con rive. Capon with rice. Croccante di mandorle. Almond sweet. Ostriche alla Napolitana. Oyster savoury.

The Ninth Day

"Since I have been associated with the production of a dinner, I have had my eyes opened as to the complicated nature of the task, and the numerous strings which have to be pulled in order to ensure success," said the Colonel; "but, seeing that a dinner-party with well-chosen sympathetic guests and distinguished dishes represents one of the consummate triumphs of civilisation, there is no reason to wonder. To achieve a triumph of any sort demands an effort."

"Effort," said Miss Macdonnell. "Yes, effort is the word I associate with so many middle-class English dinners. It is an effort to the hosts, who regard the whole business as a mere paying off of debts; and an effort to the guests, who, as they go to dress, recall grisly memories of former similar experiences. It often astonishes me that dinner-giving of this character should still flourish."

"The explanation is easy," said Van der Roet; "it flourishes because it gives a mark of distinction. It is a delicious moment for Mrs. Johnson when she is able to say to Mrs. Thompson, 'My dear, I am quite worn-out; we dined out every day last week, and have four more dinners in the next five days.' These good people show their British grit by the persistency with which they go on with their penitential hospitality, and their lack of ideas in never attempting to modify it so as to make it a pleasure instead of a disagreeable duty."

"It won't do to generalise too widely, Van der Roet," said Sir John. "Some of these good people surely enjoy their party-giving; and, from my own experience of one or two houses of this sort, I can assure you the food is quite respectable. The great imperfection seems to lie in the utter want of consideration in the choice of guests. A certain number of people and a certain quantity of food shot into a room, that is their notion of a dinner-party."

"Of course we understand that the success of a dinner depends much more on the character of the guests than on the character of the food," said Mrs. Sinclair; "and most of us, I take it, are able to fill our tables with pleasant friends; but what of the dull people who know none but dull people? What gain will they get by taking counsel how they shall fill their tables?"

"More, perhaps, than you think, dear Mrs. Sinclair," said Sir John. "Dull people often enjoy themselves immensely when they meet dull people only. The frost comes when the host unwisely mixes in one or two guests of another sort—people who give themselves airs of finding more pleasure in reading Stevenson than the sixpenny magazines, and who don't know where Hurlingham is. Then the sheep begin to segregate themselves from the goats, and the feast is manque."

"Considering what a trouble and anxiety a dinner-party must be to the hostess, even under the most favouring conditions, I am always at a loss to discover why so many women take so much pains, and spend a considerable sum of money as well, over details which are unessential, or even noxious," said Mrs. Wilding. "A few flowers on the table are all very well—one bowl in the centre is enough—but in many houses the cost of the flowers equals, if it does not outrun, the cost of all the rest of the entertainment. A few roses or chrysanthemums are perfect as accessories, but to load a table with flowers of heavy or pungent scent is an outrage. Lilies of the valley are lovely in proper surroundings, but on a dinner-table they are anathema. And then the mass of paper monstrosities which crowd every corner. Swans, nautilus shells, and even wild boars are used to hold up the menu. Once my menu was printed on a satin flag, and during the war the universal khaki invaded the dinner table. Ices are served in frilled baskets of paper, which have a tendency to dissolve and amalgamate with the sweet. The only paper on the table should be the menu, writ plain on a handsome card."

"No one can complain of papery ices here," said the Marchesa. "Ices may be innocuous, but I don't favour them, and no one seems to have felt the want of them; at least, to adopt the phrase of the London shopkeeper, 'I have had no complaints.' And even the ice, the very emblem of purity, has not escaped the touch of the dinner-table decorator. Only a few days ago I helped myself with my fingers to what looked like a lovely peach, and let it flop down into the lap of a bishop who was sitting next to me. This was the hostess's pretty taste in ices."

"They are generally made in the shape of camelias this season," said Van der Roet. "I knew a man who took one and stuck it in his buttonhole."

"I must say I enjoy an ice at dinner," said Lady Considine. "I know the doctors abuse them, but I notice they always eat them when they get the chance."

"Ah, that is merely human inconsistency," said Sir John. "I am inclined to agree with the Marchesa that ice at dinner is an incongruity, and may well be dispensed with. I think I am correct, Marchesa, in assuming that Italy, which has showered so many boons upon us, gave us also the taste for ices."

"I fear I must agree," said the Marchesa. "I now feel what a blessing it would have been for you English if you had learnt from us instead the art of cooking the admirable vegetables your gardens produce. How is it that English cookery has never found any better treatment for vegetables than to boil them quite plain? French beans so treated are tender, and of a pleasant texture on the palate, but I have never been able to find any taste in them. They are tasteless largely because the cook persists in shredding them into minute bits, and I maintain that they ought to be cooked whole—certainly when they are young—and sautez, a perfectly plain and easy process, which is hard to beat. Plain boiled cauliflower is doubtless good, but cooked alla crema it is far better; indeed, it is one of the best vegetable dishes I know. But perhaps the greatest discovery in cookery we Italians ever made was the combination of vegetables and cheese. There are a dozen excellent methods of cooking cauliflower with cheese, and one of these has come to you through France, choux-fleurs au gratin, and has become popular. Jerusalem artichokes treated in the same fashion are excellent; and the cucumber, nearly always eaten raw in England, holds a first place as a vegetable for cooking. I seem to remember that every one was loud in its praises when we tasted it as an adjunct to Manzo alla Certosina. Why is it that celery is for the most part only eaten raw with cheese? We have numberless methods of cooking it in Italy, and beetroot and lettuce as well. There is no spinach so good as English, and nowhere is it so badly cooked; it is always coarse and gritty because so little trouble is taken with it, and I can assure you that the smooth, delicate dish which we call Flano di spinacci is not produced merely by boiling and chopping it, and turning it out into a dish."

> Menu—Lunch Minestrone alla Milanese. Vegetable broth. Coniglio alla Provenzale. Rabbit alla Provenzale. Insalata di pomidoro. Tomato salad.

Menu—Dinner. Zuppa alla Maria Pia. Soup alla Maria Pia. Anguilla con ortaggi alla Milanese. Eels with vegetables. Manzo con sugo di barbabietoli. Fillet of beef with beetroot sauce. Animelle alla parmegiana. Sweetbread with parmesan. Perniciotti alla Gastalda. Partridges alla Gastalda. Uova ripiani. Stuffed eggs.

The Tenth Day

The sun rose on the tenth and last day at the "Laurestinas" as he was wont to rise on less eventful mornings. At breakfast the Marchesa proposed that the lunch that day should be a little more ornate than usual, and the dinner somewhat simpler. She requisitioned the services of six of the company to prepare the lunch, and at the same time announced that they would all have a holiday in the afternoon except Mrs. Sinclair, whom she warned to be ready to spend the afternoon in the kitchen helping prepare the last dinner.

Four dishes, all admirable, appeared at lunch, and several of the party expressed regret that the heat of the weather forbade them from tasting every one; but Sir John was not of these. He ate steadily through the menu, and when he finally laid down his knife and fork he heaved a sigh, whether of satisfaction or regret it were hard to say.

"It is a commonplace of the deepest dye to remark that ingratitude is inherent in mankind," he began; "I am compelled to utter it, however, by the sudden longing I feel for a plate from the hand of the late lamented Narcisse after I have eaten one of the best luncheons ever put on a table."

"Experience of one school of excellence has caused a hankering after the triumphs of another," said Miss Macdonnell "There is one glory of the Marchesa, there is, or was, another of Narcisse, and the taste of the Marchesa's handiwork has stimulated the desire of comparision. Never mind, Sir John, perhaps in another world Narcisse may cook you—"

"Oh stop, stop, for goodness' sake," cried Sir John, "I doubt whether even he could make me into a dainty dish to set before the King of Tartarus, though the stove would no doubt be fitted with the latest improvements and the fuel abundant."

"Really, Sir John, I'm not sure I ought not to rise and protest," said Mrs. Wilding, "and I think I would if it weren't our last day."

"Make a note of Sir John's wickedness, and pass it on to the Canon for use in a sermon," said Van der Roet.

"I can only allow you half-an-hour, Laura," said the Marchesa to Mrs. Sinclair, "then you must come and work with me for the delectation of these idle people, who are going to spend the afternoon talking scandal under the chestnuts."

"I am quite ready to join you if I can be of any help," said Mrs. Gradinger. "When knowledge is to be acquired, I am always loath to stand aside, not for my own sake so much as for the sake of others less fortunate, to whom I might possibly impart it hereafter."

"You are very good," said the Marchesa, "but I think I must adhere to my original scheme of having Mrs. Sinclair by herself. I see coffee is now being taken into the garden, so we will adjourn, if you please."

After the two workers had departed for the kitchen, an unwonted silence fell on the party under the chestnuts. Probably every one was pondering over the imminent dissolution of the company, and wondering whether to regret or rejoice. The peace had been kept marvellously well, considering the composition of the company. Mrs. Fothergill at times had made a show of posing as the beneficent patron, and Mrs. Gradinger had essayed to teach what nobody wanted to learn; but firm and judicious snubbing had kept these persons in their proper places. Nearly every one was sorry that the end had come. It had been real repose to Mrs. Wilding to pass ten days in an atmosphere entirely free from all perfume of the cathedral close. Lady Considine had been spending freely of late, and ten days' cessation of tradesmen's calls, and servants on board wages, had come as a welcome relief. Sir John had gained a respite from the task he dreaded, the task of going in quest of a successor to Narcisse. Now as he sat consuming his cigarette in the leisurely fashion so characteristic of his enjoyment—and those who knew him best were wont to say that Sir John practiced few arts so studiously as that of enjoyment—he could not banish the figure of Narcisse from his reverie. A horrible thought assailed him that this obsession might spring from the fact that on this very morning Narcisse might have taken his last brief walk out of the door of La Roquette, and that his disembodied spirit might be hovering around. Admirable as the cookery of the Marchesa had been, and fully as he had appreciated it, he felt he would give a good deal to be assured that on this the last evening of the New Decameron he might sit down to a dinner prepared by the hand of his departed chef.

That evening the guests gathered round the table with more empressement than usual. The Marchesa seemed a little flurried, and Mrs. Sinclair, in a way, shared her excitement. The menu, for the first time, was written in French, a fact which did not escape Sir John's eye. He made no remark as to the soup; it was the best of its kind, and its French name made it no better than the other triumphs in the same field which the Marchesa had achieved. But when Sir John tasted the first mouthful of the fish he paused, and after a reflective and regretful look at his plate, he cast his eye round the table. All the others, however, were too busily intent in consuming the Turbot la Vatel to heed his interrogative glance, so he followed suit, and after he had finished his portion, asked, sotto voce, for another bit.

In the interval before the service of the next dish Sir John made several vain attempts to catch the Marchesa's eye, and more than once tried to get in a word; but she kept up a forced and rather nervous conversation with Lady Considine and Van der Roet, and refused to listen. As Sir John helped himself to the next dish, Venaison sauce Grand Veneur, the feeling of astonishment which had seized him when he first tasted the fish deepened into something like Consternation. Had his palate indeed deceived him, or had the Marchesa, by some subtle effort of experimental genius, divined the secret of Narcisse—the secret of that incomparable sauce, the recipe of which was safely bestowed in his pocket-book? Occasionally he had taken a brief nap under the verandah after lunch: was it possible that in his sleep he might have murmured, in her hearing, words which gave the key of the mystery, and the description of those ingredients which often haunted his dreams? One thing was certain, that the savour which rose from the venison before him was the same which haunted his memory as the parting effort of the ill-starred Narcisse.

Sir John was the least superstitious of mortals, still here he was face to face with one of these conjunctions of affairs which the credulous accept as manifestations of some hidden power, and sceptics as coincidences and nothing more. All the afternoon he had been thinking of Narcisse, and yearning beyond measure for something suggestive of his art; and here, on his plate before him, was food which might have been touched by the vanished hand. The same subtle influence pervaded the Chartreuse a la cardinal, the roast capon and salad, and the sweet. At last, when the dinner was nearly over, and when the Marchesa had apparently said all she had to say to Van der Roet, he lifted up his voice and said, "Marchesa, who gave you the recipe for the sauce with which the venison was served this evening?"

The Marchesa glanced at Mrs. Sinclair, and then struck a handbell on the table. The door opened, and a little man, habited in a cook's dress of spotless white, entered and came forward. "M. Narcisse," said the Marchesa, "Sir John wants to know what sauce was used in dressing the venison; perhaps you can tell him."

Here the Marchesa rose and left the room, and all the rest followed her, feeling it was unmeet that such a reunion should be witnessed by other eyes, however friendly they might be.

"Now, you must tell us all about it," said Lady Considine, as soon as they got into the drawing-room, "and how you ever managed to get him out of this scrape."

"Oh, there isn't much to tell," said the Marchesa. "Narcisse was condemned, indeed, but no one ever believed he would be executed. One of my oldest friends is married to an official high up in the Ministry of Justice, and I heard from her last week that Narcisse would certainly be reprieved; but I never expected a free pardon. Indeed, he got this entirely because it was discovered that Mademoiselle Sidonie, his accomplice, was really a Miss Adah Levine, who had graduated at a music-hall in East London, and that she had announced her intention of retiring to the land of her birth, and ascending to the apex of her profession on the strength of her Parisian reputation. Then it was that the reaction in favour of Narcisse set in; the boulevards could not stand this. The journals dealt with this new outrage in their best Fashoda style; the cafes rang with it: another insult cast upon unhappy France, whose destiny was, it seemed, to weep tears of blood to the end of time. There were rumours of an interpellation in the Chamber, the position of the Minister of the Interior was spoken of as precarious, indeed the Eclaireur reported one evening that he had resigned. Pockets were picked under the eyes of sergents de ville, who were absorbed in proclaiming to each other their conviction of the innocence of Narcisse, and the guilt of cette coquine Anglaise. Cabmen en course ran down pedestrians by the dozen, as they discussed l'affaire Narcisse to an accompaniment of whip-cracking. In front of the Cafe des Automobiles a belated organ-grinder began to grind the air of Mademoiselle Sidonie's great song Bonjour Coco, whereupon the whole company rose with howls and cries of, 'A bas les Anglais, a bas les Juifs. 'Conspuez Coco.' In less than five minutes the organ was disintegrated, and the luckless minstrel flying with torn trousers down a side street. For the next few days la haute gomme promenaded with fragments of the piano organ suspended from watch chains as trophies of victory. But this was not all. Paris broke out into poetry over l'affaire Narcisse, and here is a journal sent to me by my friend which contains a poem in forty-nine stanzas by Aristophane le Beletier, the cher maitre of the 'Moribonds,' the very newest school of poetry in Paris. I won't inflict the whole of it on you, but two stanzas I must read —

> "'Puisse-je te rappeler loin des brouillards maudits. Vers la France, sainte mere et nourrice! Reviens a Lutece, de l'art vrai paradis, Je t'evoque, O Monsieur Narcisse! Quitte les saignants bifteks, de tes mains sublimes Gueris le sein meurtri de ta mere! Detourne ton glaive trenchant de tes freles victimes Vers l'Albion et sa triste Megere.'"

"Dear me, it sounds a little like some other Parisian odes I have read recently," said Lady Considine. "The triste Megere, I take it, is poor old Britannia, but what does he mean by his freles victimes?"

"No doubt they are the pigeons and the rabbits, and the chickens and the capons which Narcisse is supposed to have slaughtered in hecatombs, in order to gorge the brutal appetite of his English employer," said Miss Macdonnell. "After disregarding such an appeal as this M. Narcisse had better keep clear of Paris for the future, for if he should go back and be recognised I fancy it would be a case of 'conspuvez Narcisse.'"

"The French seem to have lost all sense of exactness," said Mrs. Gradinger, "for the lines you have just read would not pass muster as classic. In the penultimate line there are two syllables in excess of the true Alexandrine metre, and the last line seems too long by one. Neither Racine nor Voltaire would have taken such liberties with prosody. I remember a speech in Phaedre of more than a hundred lines which is an admirable example of what I mean. I dare say some of you know it. It begins:—

"Perfide! oses-tu bien te montrer devant moi? Monstre,"

but before the reciter could get fairly under way the door mercifully opened, and Sir John entered. He advanced towards the Marchesa, and shook her warmly by the hand, but said nothing; his heart was evidently yet too full to allow him to testify his relief in words. He was followed closely by the Colonel, who, taking his stand on the hearth-rug, treated the company to a few remarks, couched in a strain of unwonted eulogy. In the whole course of his life he had never passed a more pleasant ten days, though, to be sure, he had been a little mistrustful at first. As to the outcome of the experiment, if they all made even moderate use of the counsels they had received from the Marchesa, the future of cookery in England was now safe. He was not going to propose a formal vote of thanks, because anything he could say would be entirely insufficient to express the gratitude he felt, and because he deemed that each individual could best thank the Marchesa on his or her behalf.

There was a momentary silence when the Colonel ceased, and then a clearing of the throat and a preliminary movement of the arms gave warning that Mrs. Gradinger was going to speak. The unspoken passage from Racine evidently sat heavily on her chest. Abstracted and overwrought as he was, these symptoms aroused in

Sir John a consciousness of impending danger, and he rushed, incontinent, into the breach, before the lady's opening sentence was ready.

"As Colonel Trestrail has just remarked, we, all of us, are in debt to the Marchesa in no small degree; but, in my case, the debt is tenfold. I am sure you all understand why. As a slight acknowledgment of the sympathy I have received from every one here, during my late trial, I beg to ask you all to dine with me this day week, when I will try to set before you a repast a la Francaise, which I hope may equal, I cannot hope that it will excel, the dinners all'Italiana we have tasted in this happy retreat. Narcisse and I have already settled the menu."

"I am delighted to accept," said the Marchesa. "I have no engagement, and if I had I would throw my best friend over."

"And this day fortnight you must all dine with me," said Mrs. Sinclair. "I will spend the intervening days in teaching my new cook how to reproduce the Marchesa's dishes. Then, perhaps, we may be in a better position to decide on the success of the Marchesa's experiment."

The next morning witnessed the dispersal of the party. Sir John and Narcisse left by an early train, and for the next few days the reforming hand of the last-named was active in the kitchen. He arrived before the departure of the temporary aide, and had not been half-an-hour in the house before there came an outbreak which might easily have ended in the second appearance of Narcisse at the bar of justice, as homicide, this time to be dealt with by a prosaic British jury, which would probably have doomed him to the halter. Sir John listened over the balusters to the shrieks and howls of his recovered treasure, and wisely decided to lunch at his club. But the club lunch, admirable as it was, seemed flat and unappetising after the dainty yet simple dishes he had recently tasted; and the following day he set forth to search for one of those Italian restaurants, of which he had heard vague reports. Certainly the repast would not be the same as at the "Laurestinas," but it might serve for once. Alas! Sir John did not find the right place, for there are "right places" amongst the Italian restaurants of London. He beat a hasty retreat from the first he entered, when the officious proprietor assured him that he would serve up a dejeuner in the best French style. At the second he chose a dish with an Italian name, but the name was the only Italian thing about it. The experiment had failed. It seemed as if Italian restaurateurs were sworn not to cook Italian dishes, and the next day he went to do as best he could at the club.

But before he reached the club door he recalled how, many years ago, he and other young bloods used to go for chops to Morton's, a queer little house at the back of St. James' Street, and towards Morton's he now turned his steps. As he entered it, it seemed as if it was only yesterday that he was there. He beheld the waiter, with mouth all awry, through calling down the tube. The same old mahogany partitions to the boxes, and the same horse-hair benches. Sir John seated himself in a box, where there was one other luncher in the corner, deeply absorbed over a paper. This luncher raised his head and Sir John recognised Van der Roet.

"My dear Vander, whatever brought you here, where nothing is to be had but chops? I didn't know you could eat a chop."

"I didn't know it myself till to-day," said Van der Roet, with a hungry glance at the waiter, who rushed by with a plate of smoking chops in each hand. "The fact is, I've had a sort of hankering after an Italian lunch, and I went out to find one, but I didn't exactly hit on the right shop, so I came here, where I've been told you can get a chop properly cooked, if you don't mind waiting."

"Ah! I see," said Sir John, laughing. "We've both been on the same quest, and have been equally unlucky. Well, we shall satisfy our hunger here at any rate, and not unpleasantly either."

"I went to one place," said Van der Roet "and before ordering I asked the waiter if there was any garlic in the dish I had ordered. 'Garlic, aglio, no, sir, never.' Whereupon I thought I would go somewhere else. Next I entered the establishment of Baldassare Romanelli. How could a man with such a name serve anything else than the purest Italian cookery, I reasoned, so I ordered, unquestioning, a piatio with an ideal Italian name, Manzo alla Terracina. Alas! the beef used in the composition thereof must have come in a refrigerating chamber from pastures more remote than those of Terracina, and the sauce served with it was simply fried onions. In short, my dish was beefsteak and onions, and very bad at that. So in despair I fell back upon the trusty British chop."

As Van der Roet ceased speaking another guest entered the room, and he and Sir John listened attentively while the new-comer gave his order. There was no mistaking the Colonel's strident voice. "Now, look here! I want a chop underdone, underdone, you understand, with a potato, and a small glass of Scotch whisky, and I'll sit here."

"The Colonel, by Jove," said Sir John; "I expect he's been restaurant-hunting too."

"Hallo!" said the Colonel, as he recognised the other two, "I never thought I should meet you here: fact is, I've been reading about 'agricultural depression' and how it is the duty of everybody to eat chops so as to encourage the mutton trade, and that sort of thing."

"Oh, Colonel, Colonel," said Van der Roet. "You know you've been hungering after the cookery of Italy, and trying to find a genuine Italian lunch, and have failed, just as Sir John and I failed, and have come here in despair. But never mind, just wait for a year or so, until the 'Cook's Decameron' has had a fair run for its money, and then you'll find you'll fare as well at the ordinary Italian restaurant as you did at the 'Laurestinas,' and that's saying a good deal."

PART II — RECIPES

Sauces

As the three chief foundation sauces in cookery, Espagnole or brown sauce, Velute or white sauce, and Bechamel, are alluded to so often in these pages, it will be well to give simple Italian recipes for them.

Australian wines may be used in all recipes where wine is mentioned: Harvest Burgundy for red, and Chasselas for Chablis.

No. 1. Espagnole, or Brown Sauce

The chief ingredient of this useful sauce is good stock, to which add any remnants and bones of fowl or game. Butter the bottom of a stewpan with at least two ounces of butter, and in it put slices of lean veal, ham, bacon, cuttings of beef, fowl, or game trimmings, three peppercorns, mushroom trimmings, a tomato, a carrot and a turnip cut up, an onion stuck with two cloves, a bay leaf, a sprig of thyme, parsley and marjoram. Put the lid on the stewpan and braize well for fifteen minutes, then stir in a tablespoonful of flour, and pour in a quarter pint of good boiling stock and boil very gently for fifteen minutes, then strain through a tamis, skim off all the grease, pour the sauce into an earthenware vessel, and let it get cold. If it is not rich enough, add a little Liebig or glaze. Pass through a sieve again before using.

No. 2. Velute Sauce

The same as above, but use white stock, no beef, and only pheasant or fowl trimmings, button mushrooms, cream instead of glaze, and a chopped shallot.

No. 3. Bechamel Sauce

Ingredients: Butter, ham, veal, carrots, shallot, celery bay leaf, cloves, thyme, peppercorns, potato flour, cream, fowl stock.

Prepare a mirepoix by mixing two ounces of butter, trimmings of lean veal and ham, a carrot, a shallot, a little celery, all cut into dice, a bay leaf, two cloves, four peppercorns, and a little thyme. Put this on a moderate fire so as not to let it colour, and when all the moisture is absorbed add a tablespoonful of potato flour. Mix well, and gradually add equal quantities of cream and fowl stock, and stir till it boils. Then let it simmer gently. Stir occasionally, and if it gets too thick, add more cream and white stock. After two hours pass it twice slowly through a tamis so as to get the sauce very smooth.

No. 4. Mirepoix Sauce (for masking)

Ingredients: Bacon, onions, carrots, ham, a bunch of herbs, parsley, mushrooms, cloves, peppercorns, stock, Chablis.

Put the following ingredients into a stewpan: Some bits of bacon and lean ham, a carrot, all cut into dice, half an onion, a bunch of herbs, a few mushroom cuttings, two cloves, and four peppercorns. To this add one and a quarter pint of good stock and a glass of Chablis, boil rapidly for ten minutes then simmer till it is reduced to a third. Pass through a sieve and use for masking meat, fowl, fish, &c.

No. 5. Genoese Sauce

Ingredients: Onion, butter, Burgundy, mushrooms, truffles, parsley, bay leaf, Espagnole sauce (No.1), blond of veal, essence of fish, anchovy butter, crayfish or lobster butter.

Cut up a small onion and fry it in butter, add a glass of Burgundy, some cuttings of mushrooms and truffles, a pinch of chopped parsley and half a bay leaf. Reduce half. In another saucepan put two cups of Espagnole sauce, one cup of veal stock, and a tablespoonful of essence of fish, reduce one-third and add it to the other saucepan, skim off all the grease, boil for a few minutes, and pass through a sieve. Then stir it over the fire, and add half a teaspoonful of crayfish and half of anchovy butter.

No. 6. Italian Sauce

Ingredients: Chablis, mushrooms, leeks, a bunch of herbs, peppercorns, Espagnole sauce, game gravy or stock, lemon.

Put into a stewpan two glasses of Chablis, two tablespoonful of mushroom trimmings, a leek cut up, a bunch of herbs, five peppercorns, and boil till it is reduced to half. In another stewpan mix two glasses of Espagnole (No. 1) or Velute sauce (No 2) and half a glass of game gravy, boil for a few minutes then blend the contents of the two stewpans, pass through a sieve, and add the juice of a lemon.

No. 7. Ham Sauce, Salsa di Prosciutto

Ingredients: Ham, Musca or sweet port, vinegar, basil spice.

Cut up an ounce of ham and pound it in a mortar then mix it with three dessert spoonsful of port or Musca and a teaspoonful of vinegar a little dried basil and a pinch of spice. Boil it up, and then pass it through a sieve and warm it up in a bain-marie. Serve with roast meats. If you cannot get a sweet wine add half a teaspoonful of sugar. Australian Muscat is a good wine to use.

No. 8. Tarragon Sauce

Ingredients: Tarragon, stock, butter, flour.

To half a pint of good stock add two good sprays of fresh tarragon, simmer for quarter of an hour in a stewpan and keep the lid on. In another stewpan melt one ounce of butter and mix it with three dessert-spoonsful of flour, then gradually pour the stock from the first stewpan over it, but take out the tarragon. Mix well, add a teaspoonful of finely chopped tarragon and boil for two minutes.

No. 9. Tomato Sauce

Ingredients: Tomatoes, ham, onions, basil, salt, oil, garlic, spices.

Broil three tomatoes, skin them and mix them with a tablespoonful of chopped ham, half an onion, salt, a dessert-spoonful of oil, a little pounded spice and basil. Then boil and pass through a sieve. Whilst the sauce is boiling, put in a clove of garlic with a cut, but remove it before you pass the sauce through the sieve.

No. 10. Tomato Sauce Piquante

Ingredients: Ham, butter, onion, carrot, celery, bay leaf, thyme, cloves, peppercorns, vinegar, Chablis, stock, tomatoes, Velute or Espagnole sauce, castor sugar, lemon.

Cut up an ounce of ham, half an onion, half a carrot, half a stick of celery very fine, and fry them in butter together with a bay leaf, a sprig of thyme, one clove and four peppercorns. Over this pour a third of a cup of vinegar, and when the liquid is all absorbed, add half a glass of Chablis and a cup of stock. Then add six tomatoes cut up and strained of all their liquid. Cook this in a covered stewpan and pass it through a sieve, but see that none of the bay leaf or thyme goes through. Mix this sauce with an equal quantity of Velute (No. 2) or Espagnole sauce, (No. 1), let it boil and pass through a sieve again and at the last add a teaspoonful of castor sugar, the juice of half a lemon, and an ounce of fresh butter. (Another tomato sauce may be made like this, but use stock instead of vinegar and leave out the lemon juice and sugar.)

No. 11. Mushroom Sauce

Ingredients: Velute sauce, essence of mushrooms, butter.

Mix two dessert-spoonsful of essence of mushrooms with a cupful of Velute sauce (No. 2), reduce, keep on stirring, and just before serving add an ounce of butter. This sauce can be made with essence of truffle, or game, or shallot.

No. 12. Neapolitan Sauce

Ingredients: Onions, ham, butter, Marsala, blond of veal, thyme, bay leaf, peppercorns, cloves, mushrooms, Espagnole sauce (No. 1), tomato sauce, game stock or essence.

Fry an onion in butter with some bits of cut-up ham, then pour a glass of Marsala over it, and another of blond of veal, add a sprig of thyme, a bay leaf, four peppercorns, a clove, a tablespoonful of mushroom cuttings, and reduce half. In another saucepan put two cups of Espagnole sauce, one cupful of tomato sauce, and half a cup of game stock or essence. Reduce a third, and add the contents of the first saucepan, boil the sauce a few minutes, and pass it through a sieve. Warm it up in a bain-marie before using.

No. 13. Neapolitan Anchovy Sauce

Ingredients: Anchovies, fennel, flour, spices, parsley, marjoram, garlic, lemon juice, vinegar, cream.

Wash three anchovies in vinegar, bone and pound them in a mortar with a teaspoonful of chopped fennel and a pinch of cinnamon. Then mix in a teaspoonful of chopped parsley and marjoram, a squeeze of lemon juice, a teaspoonful of flour, half a gill of boiled cream and the bones of the fish for which you will use this sauce. Pass through a sieve, add a clove of garlic with a cut in it, and boil. If the fish you are using is cooked in the oven, add a little of the liquor in which it has been cooked to the sauce. Take out the garlic before serving. Instead of anchovies you may use caviar, pickled tunny, or any other pickled fish.

No. 14. Roman Sauce (Salsa Agro-dolce)

Ingredients: Espagnole sauce, stock, burnt sugar, vinegar, raisins, pine nuts or almonds.

Mix two spoonsful of burnt sugar with one of vinegar, and dilute with a little good stock. Then add two cups of Espagnole sauce (No. 1), a few stoned raisins, and a few pinocchi* (pine nuts) or shredded almonds. Keep this hot in a bain-marie, and serve with cutlets, calf's head or feet or tongue.

*The pinocchi which Italians use instead of almonds can be bought in London when in season.

No. 15. Roman Sauce (another way)

Ingredients: Espagnole sauce, an onion, butter, flour, lemon, herbs, nutmeg, raisins, pine nuts or almonds, burnt sugar.

Cut up a small bit of onion, fry it slightly in butter and a little flour, add the juice of a lemon and a little of the peel grated, a bouquet of herbs, a pinch of nutmeg, a few stoned raisins, shredded almonds or pinocchi, and a tablespoonful of burnt sugar. Add this to a good Espagnole (No. 1), and warm it up in a bain-marie.

No. 16. Supreme Sauce

Ingredients: White sauce, fowl stock, butter.

Put three-quarters of a pint of white sauce into a saucepan, and when it is nearly boiling add half a cup of concentrated fowl stock. Reduce until the sauce is quite thick, and when about to serve pass it through a tamis into a bain-marie and add two tablespoonsful of cream.

No. 17. Pasta marinate (For masking Italian Frys)

Ingredients: Semolina flour, eggs, salt, butter (or olive oil), vinegar, water.

Mix the following ingredients well together: two ounces of semolina flour, the yolks of two eggs, a little salt, and two ounces of melted butter. Add a glass of water so as to form a liquid substance. At the last add the whites of two eggs beaten up to a snow. This will make a good paste for masking meat, fish, vegetables, or sweets which are to be fried in the Italian manner, but if for meat or vegetables add a few drops of vinegar or a little lemon juice.

No. 18. White Villeroy

Ingredients: Butter, flour, eggs, cream, nutmeg, white stock.

Make a light-coloured roux by frying two ounces of butter and two ounces of flour, stir in some white stock and keep it very smooth. Let it boil, and add the yolks of three eggs, mixed with two tablespoonsful of cream and a pinch of nutmeg. Pass it through a sieve and use for masking cutlets, fish, &c.

Soups

No. 19. Clear Soup

Ingredients: Stock meat, water, a bunch of herbs (thyme, parsley, chervil, bay leaf, basil, marjoram), three carrots, three turnips, three onions, three cloves stuck in the onions, one blade of mace.

Cut up three pounds of stock meat small and put it in a stock pot with two quarts of cold water, three carrots, and three turnips cut up, three onions with a clove stuck in each one, a bunch of herbs and a blade of mace. Let it come to the boil and then draw it off, at once skim off all the scum, and keep it gently simmering, and occasionally add two or three tablespoonsful of cold water. Let it simmer all day, and then strain it through a fine cloth.

Some of the liquor in which a calf's head has been cooked, or even a calf's foot, will greatly improve a clear soup.

The stock should never be allowed to boil as long as the meat and vegetables are in the stock pot.

No. 20. Zuppa Primaverile (Spring Soup)

Ingredients: Clear soup, vegetables.

Any fresh spring vegetables will do for this soup, but they must all be cooked separately and put into the soup at the last minute. It is best made with fresh peas, asparagus tips, and a few strips of tarragon.

No. 21. Soup alla Lombarda

Ingredients: Clear soup, fowl forcemeat, Bechamel (No. 3), peas, lobster butter, eggs, asparagus.

Make a firm forcemeat of fowl and divide it into three parts, to the first add two spoonsful of cream Bechamel, to the second four spoonsful of puree of green peas, to the third two spoonsful of lobster butter and the yolk of an egg; thus you will have the Italian colours, red, white, and green. Butter a pie dish and make little quenelles of the forcemeat. Just before serving boil them for four minutes in boiling stock, take them out carefully and put them in a warm soup tureen with two spoonsful of cooked green peas and pour a very fresh clear soup over them. Hand little croutons fried in lobster butter separately.

No. 22. Tuscan Soup

Ingredients: Stock, eggs.

Whip up three or four eggs, gradually add good stock to them, and keep on whisking them up until they begin to curdle. Keep the soup hot in a bain-marie.

No. 23. Venetian Soup

Ingredients: Clear soup, butter, flour, Parmesan, eggs.

Make a roux by frying two ounces of butter and two ounces of flour, add an ounce of grated cheese and half a cup of good stock. Mix up well so as to form a paste, and then take it off the fire and add the yolks of four eggs, mix again and form the again and form the paste into little quenelles. Boil these in a little soup, strain off, put them into the tureen and pour a good clear soup over them.

No. 24. Roman Soup

Ingredients: Stock, butter, eggs, salt, crumb of bread, parsley, nutmeg, flour, Parmesan.

Mix three and a half ounces of butter with two eggs and four ounces of crumbs of bread soaked in stock, a little chopped parsley, salt, and a pinch of nutmeg. Reduce this and add two tablespoonsful of flour and one of grated Parmesan. Form this into little quenelles and boil them in stock for a few minutes put them into a tureen and pour a good clear soup over them.

No. 25. Soup alla Nazionale

Ingredients: Clear soup, savoury custard.

Make a savoury custard and divide it into three parts, one to be left white, another coloured red with tomato, and the third green with spinach. Put a layer of each in a buttered saucepan and cook for about ten minutes, cut it into dice, so that you have the three Italian colours (red, white, and green) together, then put the custard into a soup tureen and pour a good clear soup over it.

No. 26. Soup alla Modanese

Ingredients: Stock, spinach, butter, salt, eggs, Parmesan, nutmeg, croutons.

Wash one pound of spinach in five or six waters, then chop it very fine and mix it with three ounces of butter, salt it and warm it up. Then let it get cold, pass through a hair sieve, and add two eggs, a tablespoonful of grated Parmesan, and very little nutmeg. Add this to some boiling stock in a copper saucepan, put on the lid, and on the top put some hot coals so that the eggs may curdle and help to thicken the soup. Serve with fried croutons.

No. 27. Crotopo Soup

Ingredients: Clear soup, veal, ham, eggs, salt, pepper, nutmeg, rolls.

Pound half a pound of lean veal in a mortar, then add three ounces of cooked ham with some fat in it, the yolk of an egg, salt, pepper, and very little nutmeg. Pass through a sieve, cut some small French rolls into slices, spread them with the above mixture, and colour them in the oven. Then cut them in halves or quarters, put them into a tureen, and just before serving pour a very good clear soup over them.

No. 28. Soup all'Imperatrice

Ingredients: Breast of fowl, eggs, salt, pepper, ground rice, nutmeg, clear stock.

Pound the breast of a fowl in a mortar, and add to it a teaspoonful of ground rice, the yolk of an egg, salt, pepper, and a pinch of nutmeg. Pass this through a sieve, form quenelles with it, and pour a good clear soup over them.

No. 29. Neapolitan Soup

Ingredients: Fowl, potato flour, eggs, Bechamel sauce, peas, asparagus, spinach, clear soup.

Mix a quarter pound of forcemeat of fowl with a tablespoonful of potato flour, a tablespoonful of Bechamel sauce (No. 3), and the yolk of an egg; put this into a tube about the size round of an ordinary macaroni; twenty minutes before serving squirt the forcemeat into a saucepan with boiling stock, and nip off the forcemeat as it comes through the pipe into pieces about an inch and a half long. Let it simmer, and add boiled peas and asparagus tips. If you like to have the fowl macaroni white and green, you can colour half the forcemeat with a spoonful of spinach colouring. Serve in a good clear soup.

No. 30. Soup with Risotto

Ingredients: Risotto (No. 189), eggs, bread crumbs, clear or brown soup.

If you have some good risotto left, you can use it up by making it into little balls the size of small nuts. Egg and bread crumb and fry them in butter; dry them and put them into a soup tureen with hot soup. The soup may be either clear or brown.

No. 31. Soup alla Canavese

Ingredients: White stock, butter, onions, carrot, celery, tomato, cauliflower, fat bacon, parsley, sage, Parmesan, salt, pepper.

Chop up half an onion, half a carrot, half a stick of celery, a small bit of fat bacon, and fry them in two ounces of butter. Then cover them with good white stock, boil for a few minutes, pass through a sieve, and add two tablespoonsful of tomato puree. Then blanch half a cauliflower in salted water, let it get cold, drain all the water out of it, and break it up into little bunches and put them into a stock pot with the stock, a small leaf of dried sage, crumbled up, and a little chopped parsley, and let it all boil; add a pinch of grated cheese and some pepper. Serve with grated Parmesan handed separately.

No. 32. Soup alla Maria Pia

Ingredients: White stock, eggs, butter, peas, white beans, carrot, onion, leeks, celery, cream croutons.

Soak one pound of white beans for twelve hours, then put them into a stock pot with a little salt, butter, and water, add a carrot, an onion, two leeks, and a stick of celery, and simmer until the vegetables are well cooked; then take out all the fresh vegetables, drain the beans and pass them through a sieve, but first dilute them with good stock. Put this puree into a stock pot with good white stock, and when it has boiled keep it hot in a bain-marie until you are about to serve; then mix the yolk of three eggs in a cup of cream, and add this to the soup. Pour the soup into a warm tureen, add some boiled green peas, and serve with fried croutons handed separately.

No. 33. Zuppa d' Erbe (Lettuce Soup)

Ingredients: Stock, sorrel, endive, lettuce, chervil, celery, carrot, onion, French roll, Parmesan cheese.

Boil the following vegetables and herbs in very good stock for an hour: Two small bunches of sorrel, a bunch of endive, a lettuce, a small bunch of chervil, a stick of celery, a carrot and an onion, all well washed and cut up. Then put some slices of toasted French roll into a tureen and pour the above soup over them. Serve with grated Parmesan handed separately.

No. 34. Zuppa Regina di Riso (Queen's Soup)

Ingredients: Fowl stock, ground rice, milk, butter.

Put a tablespoonful of ground rice into a saucepan and gradually add half a pint of milk, boil it gently for twelve minutes in a bain-marie, but stir the whole time, so as to get it very smooth. Just before serving add an ounce of butter, pass it through a sieve, and mix it with good fowl stock.

Minestre

Minestra is a thick broth, very much like hotch-potch, only thicker. In Italy it is often served at the beginning of dinner instead of soup; it also makes an excellent lunch dish. Two or three tablespoonsful of No. 35 will be found a great improvement to any of these minestre.

No. 35. A Condiment for Seasoning Minestre, &c.

Ingredients: Onions, celery, carrots, butter, salt, stock, tomatoes, mushrooms.

Cut up an onion, a stick of celery, and a carrot; fry them in butter and salt; add a few bits of cooked ham and veal cut up, two mushrooms, and the pulp of a tomato. Cook for a quarter of an hour, and add a little stock occasionally to keep it moist. Pass through a sieve, and use for seasoning minestre, macaroni, rice, &c. It should be added when the dish is nearly cooked.

No. 36. Minestra alla Casalinga

Ingredients: Rice, butter, stock, vegetables.

All sorts of vegetables will serve for this dish. Blanch them in boiling salted water, then drain and fry them in butter. Add plenty of good stock, and put them on a slow fire. Boil four ounces of rice in stock, and when it is well done add the stock with the vegetables. Season with two or three spoonsful of No. 35, and serve with grated cheese handed separately.

No. 37. Minestra of Rice and Turnips

Ingredients: Rice, turnips, butter, gravy, tomatoes.

Cut three or four young turnips into slices and put them on a dish, strew a little salt over them, cover them with another dish, and let them stand for about two hours until the water has run out of them. Then drain the slices, put them in a frying-pan and fry them slightly in butter. Add some good gravy and mashed-up tomatoes, and after having cooked this for a few minutes pour it into good boiling stock. Add three ounces of well-washed rice, and boil for half-an-hour.

Minestra loses its flavour if it is boiled too long. In Lombardy, however, rice, macaroni, &c., are rarely boiled enough for English tastes.

No. 38. Minestra alla Capucina

Ingredients: Rice, anchovies, butter, stock, and onions.

Scale an anchovy, pound it, and fry it in butter together with a small onion cut across, and four ounces of boiled rice. Add a little salt, and when the rice is a golden brown, take out the onion and gradually add some good stock until the dish is of the consistency of rice pudding.

No. 39. Minestra of Semolina

Ingredients: Stock, semolina, Parmesan.

Put as much stock as you require into a saucepan, and when it begins to boil add semolina very gradually, and stir to keep it from getting lumpy Cook it until the semolina is soft, and serve with grated Parmesan handed separately. To one quart of soup use three ounces of semolina.

No. 40. Minestrone alla Milanese

Ingredients: Rice or macaroni, ham, bacon, stock, all sorts of vegetables.

Minestrone is a favourite dish in Lombardy when vegetables are plentiful. Boil all sorts of vegetables in stock, and add bits of bacon, ham, onions braized in butter, chopped parsley, a clove of garlic with two cuts, and rice or macaroni. Put in those vegetables first which require most cooking, and do not make the broth too thin. Leave the garlic in for a quarter of an hour only.

No. 41. Minestra of Rice and Cabbage

Ingredients: Rice, cabbage, stock, ham, tomato sauce.

Cut off the stalk and all the hard outside leaves of a cabbage, wash it and cut it up, but not too small, then drain and cook it in good stock and add two ounces of boiled rice. This minestre is improved by adding a little chopped ham and a few spoonsful of tomato sauce.

No. 42. Minestra of Rice and Celery

Ingredients: Celery, rice, stock.

Cut up a head of celery and remove all the green parts, then boil it in good stock and add two ounces of rice, and boil till it is well cooked.

Fish

No. 43. Anguilla alla Milanese (Eels).

Ingredients: Eels, butter, flour, stock, bay leaves, salt, pepper, Chablis, a macedoine of vegetables.

Cut up a big eel and fry it in two ounces of butter, and when it is a good colour add a tablespoonful of flour, about half a pint of stock, a glass of Chablis, a bay leaf, pepper, and salt, and boil till it is well cooked. In the meantime boil separately all sorts of vegetables, such as carrots, cauliflower, celery, beans, tomatoes, &c. Take out the pieces of eel, but keep them hot, whilst you pass the liquor which forms the sauce through a sieve and add the vegetables to this. Let them boil a little longer and arrange them in a dish; place the pieces of eel on them and cover with the sauce. It is most important that the eels should be served very hot.

Any sort of fish will do as well for this dish.

No. 44. Filletti di Pesce alla Villeroy (Fillets of Fish)

Ingredients: Fish, flour, butter, Villeroy.

Any sort of fish will do, turbot, sole, trout, &c. Cut it into fillets, flour them over and cook them in butter in a covered stewpan; then make a Villeroy (No. 18), dip the fillets into it and fry them in clarified butter.

No. 45. Astachi all'Italiana (Lobster)

Ingredients: Lobsters, Velute sauce, Marsala, butter, forcemeat of fish, olives, anchovy butter, button mushrooms, truffles, lemon, crayfish, Italian sauce.

Two boiled lobsters are necessary. Cut all the flesh of one of the lobsters into fillets and put them into a saucepan with half a cup of Velute sauce (No. 2) and half a glass of Marsala, and boil for a few minutes. Put a crouton of fried bread on an oval dish and cover it with a forcemeat of fish, and on this place the whole lobster, cover it with buttered paper, and put it in a moderate oven just long enough to cook the forcemeat. Then make some quenelles of anchovy butter, olives, and button mushrooms, mix them with Italian sauce (No. 6), and garnish the dish with them, and round the crouton arrange the fillets of lobster with a garnish of slices of truffle. Add a dessertspoonful of crayfish butter and a good squeeze of lemon juice to the sauce, and serve.

No. 46. Baccala alla Giardiniera (Cod)

Ingredients: Cod or hake, carrots, turnips, butter, herbs.

Boil a piece of cod or hake and break it up into flakes, then cut up two carrots and a turnip; boil them gently, and when they are half boiled drain and put them into a stewpan with an ounce of butter, half a teacup of boiling water, salt, and herbs. When they are well cooked add the fish and serve. Fillets of lemon soles may also be cooked this way.

No. 47. Triglie alla Marinara (Mullet)

Ingredients: Mullet, salt, pepper, onions, parsley, oil, water.

Cut a mullet into pieces and put it into a stewpan (with the lid on), with salt, pepper, a cut-up onion, some chopped parsley, half a wineglass of the finest olive oil and half a pint of water, and in this cook the fish gently. Arrange the fillets on a dish, pour a little of the broth over them, and add the onion and parsley. Instead of mullet you can use cod, hake, whiting, lemon sole, &c.

No. 48. Mullet alla Tolosa

Ingredients: Mullet, butter, salt, onions, parsley, almonds, anchovies, button mushrooms, tomatoes.

Cut off the fins and gills of a mullet, put it in a fireproof dish with two ounces of butter and salt. Cut up a small bit of onion, a sprig of parsley, a few blanched almonds, one anchovy, and a few button mushrooms, previously softened in hot water, and put them over the fish and bake for twenty minutes Then add two tablespoonsful of tomato sauce or puree, and when cooked serve. If you like, use sole instead of mullet.

No. 49. Mullet alla Triestina

Ingredients: Mullet (or sole or turbot), butter, salt half a lemon, Chablis.

Put the fish in a fireproof dish with one and a half ounces of butter, salt, a squeeze of lemon juice, and half a glass of Chablis. Put it on a very, slow fire and turn the fish when necessary. When it is cooked serve in the dish.

No. 50. Whiting alla Genovese

Ingredients: Whiting, butter, pepper, salt, bay leaf claret, parsley, onions, garlic capers, vinegar, Espagnole sauce, mushrooms, anchovies.

Put one or two whiting into a stewpan with two ounces of butter, salt, pepper, two bay leaves, and a glass of claret or Burgundy; cook on a hot fire and turn the fish when necessary. Have ready beforehand a remoulade sauce made in the following manner: Put in a saucepan 1 1/2 ounces of butter, half a teaspoonful of chopped parsley, half an onion, a clove of garlic (with one cut), four capers, one anchovy, all chopped up except the garlic. Then add three tablespoonsful of vinegar and reduce the sauce. Add two glasses of Espagnole sauce (No. 1) and a little good stock; boil it all up (take out the garlic and bay leaves) and pass through a sieve, then pour it over the whiting. Boil it all again for a few minutes, and before serving garnish with a few button mushrooms cooked separately. The remoulade sauce will be much better if made some hours beforehand.

No. 51. Merluzzo in Bianco (Cod)

Ingredients: Cod or whiting, salt, onions, parsley, cloves, turnips, marjoram, chervil, milk.

Boil gently in a big cupful of salted water two onions, one turnip, a pinch of chopped parsley, chervil, and marjoram and four cloves. After half an hour pass this through a sieve (but first take out the cloves), and add an equal quantity of milk and a little cream, and in this cook the fish and serve with the sauce over it.

No. 52. Merluzzo in Salamoia (Cod)

Ingredients: Cod, hake, whiting or red mullet, onions, parsley, mint, marjoram, turnips, mushrooms, chervil, cloves, salt, milk, cream, eggs.

Put a salt-spoonful of salt, two onions, a little parsley, marjoram, mint, chervil, a turnip, a mushroom, and the heads of two cloves into a stewpan and simmer in a cupful of milk for half an hour, then let all the ingredients settle at the bottom, and pass the broth through a hair sieve, and add to it an equal quantity of milk or cream, and in it cook your fish on a slow fire. When the fish is quite cooked, pour off the sauce, but leave a little on the fish to keep it warm; reduce the rest in a bain-marie; stir all the time, so that the milk may not curdle. Thicken the sauce with the yolk of an egg, and when about to serve pour it over the fish.

No. 53. Baccala in Istufato (Haddock)

Ingredients: Haddock or lemon sole, carrots, anchovies, lemon, pepper, butter, onions, flour, white wine, stock.

Stuff a haddock (or filleted lemon sole) with some slices of carrot which have been masked with a paste made of pounded anchovies, very little chopped lemon peel, salt and pepper. Then fry an onion with two cuts across it in butter. Take out the onion as soon as it has become a golden colour, flour the fish and put it in the butter, and when it has been well fried on both sides pour a glass of Marsala over it, and when it is all absorbed add a cup of fowl or veal stock and let it simmer for half an hour, then skim and reduce the sauce, pour it over the fish and serve.

No. 54. Naselli con Piselli (Whiting)

Ingredients: Whiting, onions, parsley, peas, tomatoes, butter, Parmesan, Bechamel sauce.

Cut a big whiting into two or three pieces and fry them slightly in butter, add a small bit of onion, a teaspoonful of chopped parsley and fry for a few minutes more. Then add some peas which have been cooked in salted water, three tablespoonsful of Bechamel sauce (No. 3), and three of tomato puree, and cook all together on a moderate fire.

No. 55. Ostriche alla Livornese (Oysters)

Ingredients: Oysters, parsley, shallot, anchovies, fennel pepper, bread crumbs, cream, lemon.

Detach the oysters from their shells and put then into china shells with their own liquor. Have ready a dessert-spoonful of parsley, shallot, anchovy and very little fennel, add a tablespoonful of bread crumbs and a little pepper, and mix the whole with a little cream. Put some of this mixture on each oyster, and then bake them in a moderate fire for a quarter of an hour. At the last minute add a squeeze of lemon juice to each oyster and serve on a folded napkin.

No. 56. Ostriche alla Napolitana (Oysters)

Ingredients: Oysters, parsley, celery, thyme, pepper, garlic, oil, lemon.

Prepare the oysters as above, but rub each shell with a little garlic. Put on each oyster a mixture made of chopped parsley, a little thyme, pepper, and bread crumbs. Then pour a few drops of oil on each shell, put them on the gridiron on an open fire, grill for a few minutes, and add a little lemon juice before serving.

No. 57. Ostriche alla Veneziana (Oysters)

Ingredients: Oysters, butter, shallots, truffles, lemon juice, forcemeat of fish.

Take several oysters out of their shells and cook them in butter, a little chopped shallot, and their own liquor, add a little lemon juice and then put in each of the deeper shells a layer of forcemeat made of fish and chopped truffles, then an oyster or two, and over this again another layer of the forcemeat, cover up with the top shell and put them in a fish kettle and steam them. Then remove the top shell and arrange the shells with the oysters on a napkin and serve.

No. 58. Pesci diversi alla Casalinga (Fish)

Ingredients: Any sort of fish, celery, parsley, carrots, garlic, onion, anchovies, almonds, capers, mushrooms, butter, salt, pepper, flour, tomatoes.

Chop up a stick of celery, a sprig of parsley, a carrot, an onion. Pound up an anchovy in brine (well cleaned, boned, and scaled), four shredded almonds, three capers and two mushrooms. Put all this into a saucepan with one ounce of butter, salt and pepper, and fry for a few minutes, then add a few spoonsful of hot water and a tablespoonful of flour and boil gently for ten minutes, put in the fish and cook it until it is done. If you like, you may add a little tomato sauce.

No. 59. Pesce alla Genovese (Sole or Turbot)

Ingredients: Fish (sole, mullet, or turbot), butter, salt, onion, garlic, carrots, celery, parsley, nutmeg, pepper, spice, mushrooms, tomatoes, flour, anchovies.

Fry an onion slightly in one and a half ounces of butter, add a small cut-up carrot, half a stick of celery, a sprig of parsley, and a salt anchovy (scaled), which will dissolve in the butter. Into this put the fish cut up in pieces, a pinch of spice and pepper, and let it simmer for a few minutes, then add two cut-up mushrooms, a tomato mashed up, and a little flour. Mix all together, and cook for twenty minutes.

No. 60. Sogliole in Zimino (Sole)

Ingredients: Sole, onion, beetroot, butter, celery, tomato sauce or white wine.

Cut up a small onion and fry it slightly in one ounce of butter, then add some slices of beetroot (well-washed and drained), and a little celery cut up; to this add fillets of sole or haddock, salt and pepper. Boil on a moderate on the fish kettle. When the beetroot is nearly cooked add two tablespoonsful of tomato puree and boil till all is well cooked. Instead of the tomato you may use half a glass of Chablis.

No. 61. Sogliole al tegame (Sole)

Ingredients: Sole (or mullet), butter, anchovies, parsley, garlic, capers, eggs.

Put an ounce of butter and an anchovy in a saucepan together with a sole or mullet. Fry lightly for a few minutes, then strew a little pepper and chopped parsley over it, put in a clove of garlic with one cut, and cook for half an hour, but turn the fish over when one side is sufficiently done. A few minutes before taking it off the fire add three capers and stir in the yolk of an egg at the last minute. Do not leave the garlic in more than five minutes.

No. 62. Sogliole alla Livornese (Sole)

Ingredients: Sole, butter, garlic, pepper, salt, tomatoes, fennel.

Fillet a sole and put it in a saute-pan with one and a half ounces of butter and a clove of garlic with one cut in it, then sprinkle over it a little chopped fennel, salt and pepper, and let it cook for a few minutes. Turn over the fillets when they are sufficiently cooked on one side, take out the garlic and cover the fish with a puree of tomatoes at the last.

No. 63. Sogliole alla Veneziana (Sole)

Ingredients: Sole, anchovies, butter, bacon, onion, stock, Chablis, salt, nutmeg, parsley, Spanish olives, one bay leaf.

Fillet a sole and interlard each piece with a bit of anchovy. Tie up the fillets and put them in a saute-pan with two ounces of butter, a slice of bacon or ham, and a few small slices of onion. Cover half over with good stock and a glass of Chablis, and add salt, a pinch of nutmeg, a bunch of parsley, and a bay leaf. Cover with buttered paper, and cook on a slow fire for about an hour. Drain the fish, pass the liquor through a sieve, reduce it to the consistency of a thick sauce, and pour it over the fish. Garnish each fillet with a Spanish olive stuffed with anchovy.

No. 64. Sogliole alla Parmigiana (Sole).*

Ingredients: Sole, Parmesan, butter, cream, cayenne.

Fillet a sole and wipe each piece with a clean cloth, then place them in a fireproof dish, and put a small piece of butter on each fillet. Then make a good white sauce, and mix it with two tablespoonsful of grated Parmesan and half a gill of cream. Cover the fish well with the sauce, and bake in a moderate oven for twenty minutes.

*Lemon soles may be used in any of the above-named dishes.

No. 65. Salmone alla Genovese (Salmon)

Ingredients: Salmon, Genoese sauce (No. 5), butter, lemon.

Boil a bit of salmon, drain it, take off the skin, and mask it with a Genoese sauce, to which add a spoonful of the water in which the salmon has been boiled, and at the last add a pat of fresh butter and a squeeze of lemon juice.

No. 66. Salmone alla Perigo (Salmon)

Ingredients: Salmon, forcemeat of fish, truffles, butter, Madeira, croutons of bread, crayfish tails, anchovy butter.

Cut a bit of salmon into well shaped fillets, and marinate them in lemon juice and a bunch of herbs for two hours, wipe them, put a layer of forcemeat of fish over each, and decorate them with slices of truffle. When put them into a well-buttered saute-pan with half a cup of stock and a glass of Madeira or Marsala, cover with buttered paper, and put them into a moderate oven for twenty minutes. Arrange the fillets in a circle on croutons of bread, garnish the centre with crayfish tails and with truffles cut into dice, a quarter of a pint of Velute sauce (No. 2), and half a teaspoonful of anchovy butter. Glaze the fillets and serve.

No. 67. Salmone alla giardiniera (Salmon)

Ingredients: Salmon, forcemeat of fish, vegetables, butter, Bechamel, and Espagnole sauce.

Prepare the fillets as above (No. 66), and put on each a layer of white forcemeat of fish. Cook a macedoine of vegetables separately, and garnish each fillet with some of it, then cook them in a covered stewpan Put a crouton of bread in an entree dish and garnish it with cooked peas, mixed with Bechamel sauce (No. 3), stock, and butter. Around this place the fillets of fish, leaving the centre with the peas uncovered. Pour some rich Espagnole sauce (No. 1) round the fillets and serve.

No. 68. Salmone alla Farnese (Salmon)

Ingredients: Salmon, oil, lemon juice, thyme, salt, pepper, nutmeg, mayonnaise sauce, lobster butter, gelatine, Velute sauce, olives, anchovy butter, white truffles, mushrooms in oil, crayfish.

Boil a piece of salmon, and when cold cut it into fillets and marinate them for two hours in oil, lemon juice, salt, thyme pepper, and nutmeg. Then make a good mayonnaise and add to it some lobster butter mixed with a little dissolved gelatine and Velute sauce (No. 2). Wipe the fillets and arrange them in a circle on a dish, and pour the mayonnaise over them. Then decorate the border of the dish with aspic jelly, and in the centre put some stoned Spanish olives stuffed with anchovy butter, truffles, mushrooms in oil, and crayfish tails.

No. 69. Salmone alla Santa Fiorentina (Salmon)

Ingredients: Salmon, eggs, mayonnaise, parsley, flour.

Marinate a piece of boiled salmon for an hour; take out the bone and cut the fish into fillets, wipe them, roll them in flour and dip them in eggs beaten up or in mayonnaise sauce, and fry them a good colour. Arrange in a circle on the dish, garnish with fried parsley, and serve with Dutch or mayonnaise sauce. Any fillets of fish may be cooked in this manner.

No. 70. Salmone alla Francesca (Salmon)

Ingredients: Salmon, butter, onions, parsley, salt, pepper, nutmeg, stock, Chablis, Espagnole sauce (No.1) mushrooms, anchovy butter, lemon.

Put a firm piece of salmon in a stewpan with one and a half ounces of butter, an onion cut up, a teaspoonful of chopped parsley (blanched), salt, pepper, very little nutmeg, a cup of stock, and a glass of Chablis. Cook for half an hour over a hot fire, turn the salmon occasionally, and if it gets dry, add a cup of Espagnole sauce. Let it boil until sufficiently cooked, and then put it on a dish. Into the sauce put four mushrooms cooked in white sauce, half a teaspoonful of anchovy butter and a little lemon juice. Pour the sauce over the salmon and serve.

No. 71. Fillets of Salmon in Papiliotte

Ingredients: Salmon, oil, lemon juice, salt, pepper, nutmeg, herbs.

Cut a piece of salmon into fillets, marinate them in oil, lemon juice, salt, pepper, nutmeg, and herbs for two hours. Wipe and put them into paper souffle cases with a little oil, butter, and herbs. Cook them on a gridiron, and serve with a sauce piquante made in the following manner: Half a pint of rich Espagnole sauce (No. 1) and a dessert-spoonful of New Century{*} sauce, warmed up in a bain-marie.

*Can be obtained at Messrs Lazenby's, Wigmoree Street, W.

Beef, Mutton, Veal, Lamb, &C.

No. 72. Manzo alla Certosina (Fillet of Beef)

Ingredients: Fillet of beef or rump steak, bacon, olive oil, salt, nutmeg, anchovies, herbs, stock, garlic.

Put a piece of very tender rump steak or fillet of beef into a stewpan with two slices of fat bacon and three teaspoonsful of the finest olive oil; season with salt and a tiny pinch of nutmeg; let it cook uncovered, and turn the meat over occasionally. When it is nicely browned add an anchovy minced and mixed with chopped herbs, and a small clove of garlic with one cut across it. Then cover the whole with good stock, put the cover on the stewpan, and when it is all sufficiently cooked, skim the grease off the sauce, pass it through a sieve, and pour it over the beef. Leave the garlic in for five minutes only.

No. 73. Stufato alla Florentina (Stewed Beef)

Ingredients: Beef, mutton, or veal, onions, rosemary, Burgundy, tomatoes, stock, potatoes, butter, garlic.

Cut up an onion and three leaves of rosemary, fry them slightly in an ounce of butter, then add meat (beef, mutton, or veal), cut into fair-sized pieces, salt it and fry it a little, then pour half a glass of Burgundy over it, and add two tablespoonsful of tomato conserve, or better still, fresh tomatoes in a puree. Cover up the stewpan and cook gently, stir occasionally, and add some stock if the stew gets too dry. If you like to add potatoes, cut them up, put them in the stewpan an hour before serving, and cook them with the meat. A clove of garlic with one cut may be added for five minutes.

No. 74. Coscia di Manzo al Forno (Rump Steak)

Ingredients: Rump steak, ham, salt, pepper, spice, fat bacon, onion, stock, white wine.

Lard a bit of good rump steak with bits of lean ham, and season it with salt, pepper, and a little spice, slightly brown it in butter for a few minutes, then cover it with three or four slices of fat bacon and put it into a stewpan with an onion chopped up, a cup of good stock, and half a glass of white wine; cook with the cover on the stewpan for about an hour. You may add a clove of garlic for ten minutes.

No. 75. Polpettine alla Salsa Piccante (Beef Olives)

Ingredients: Beef steak, butter, onions, stock, sausage meat.

Cut some thin slices of beef steak, and on each place a little forcemeat of fowl or veal, to which add a little sausage meat: roll up the slices of beef and cook them with butter and onions, and when they are well browned pour some stock over them, and let them absorb it. Serve with a tomato sauce (No. 10), or sauce piquante made with a quarter of a pint of rich Espagnole (No. 1), and a dessert-spoonful of New Century sauce (see No. 71 note).

No. 76. Stufato alla Milanese (Stewed Beef)

Ingredients: Rump steak, bacon, ham, salt, pepper, cinnamon, cloves, butter, onions, Burgundy.

Beat a piece of rump steak to make it tender and lard it well, cut up some bits of fat bacon and dust them over with salt, pepper, and a tiny pinch of cinnamon, and put them on the steak. Stick three cloves into the steak, then put it into a stewpan, add a little of the fat of the beef chopped up, an ounce of butter, an onion cut up, and some bits of lean ham. Put in sufficient stock to cover the steak, add a glass of Burgundy, and stew gently until it is cooked.

No. 77. Manzo Marinato Arrosto (Marinated Beef)

Ingredients: Beef, salt, larding bacon, Burgundy, vinegar, spices, herbs, flour.

Beat a piece of rump steak, or fillet to make it tender; sprinkle it well with salt and some chopped herbs, and leave it for an hour; then lard it and marinate it as follows: Half a pint of red wine (Australian Harvest Burgundy is best), half a glass of vinegar, a pinch of spice, and a bouquet of herbs; leave it in this for twenty-four hours then take it out, drain it well sprinkle it with flour, and roast it for twenty minutes before a clear fire, braize it till quite tender, then press and glaze it. The thin end of a sirloin is excellent cooked this way. Serve cold.

No. 78. Manzo con sugo di Barbabietole (Fillet of Beef)

Ingredients: Beef, beetroot, salt.

Cut up three raw beetroots put them into an earthen ware pot and cover them with water. Keep them in some warm place, and allow them to ferment for five, six, or eight days according to the season; the froth at the top of the water will indicate the necessary fermentation. The take out the pieces of beetroot, skim off all the froth, and into the fermented liquor put a good piece of tender rump steak or fillet with some salt. Braize for four hours and serve.

No. 79. Manzo in Insalata (Marinated Beef)

Ingredients: Beef, oil, salt, pepper, vinegar, parsley, capers, mushrooms, olives, vegetables.

Cook a fillet of beef (or the thin end of a sirloin), which has been previously marinated for two days in oil, salt, pepper, vinegar, and chopped parsley. When cold press and glaze it, garnish it with capers, mushrooms preserved in vinegar or gherkins, olives, and any kind of vegetables marinated like the beef. Serve cold.

No. 80. Filetto di Bue con Pistacchi (Fillets of Beef with Pistacchios)

Ingredients: Fillet of beef, oil, salt, flour, pistacchio nuts, gravy.

Cut a piece of tender beef into little fillets, and put a them in a stewpan with a tablespoonful of olive oil and salt. After they have cooked for a few minutes, powder them with flour, and strew over each fillet some chopped pistacchio nuts. Add a few spoonsful of very good boiling gravy, and cook for another half-hour.

No. 81. Scalopini di Riso (Beef with Risotto)

Ingredients: Rump steak, butter, rice, truffles, tongue, stock, mushrooms.

Slightly stew a bit of rump steak with bits of tongue and mushrooms; let it get cold, and cut it into scallops. Butter a pie dish, and garnish the bottom of it with cooked tongue and slices of cooked truffle, then over this put a layer of well-cooked and seasoned risotto (No. 190), then a layer of the scallops of beef, and then another layer of risotto. Heat in a bain-marie, and turn out of the pie dish, and serve with a very good sauce poured round it.

No. 82. Tenerumi alla Piemontese (Tendons of Veal)

Ingredients: Tendons of veal, fowl forcemeat, truffles, risotto (No. 190), a cock's comb, tongue.

Tendons of veal are that part of the breast which lies near the ribs, and forms an opaque gristly substance. Partly braize a fine bit of this joint, and press it between two plates till cold. Cut it up into fillets, and on each spread a thin layer of fowl forcemeat, and decorate with slices of truffle. Put the fillets into a stewpan, cover them with very good stock, and boil till the forcemeat and truffles are quite cooked. Prepare a risotto all'Italiana (No. 190), put it on a dish and decorate it with bits of red tongue cut into shapes, and in the centre put a whole cooked truffle and a white cock's comb, both on a silver skewer. Place the tendons of veal round the dish. Add a good Espagnole sauce (No. 1) and serve.

If you like, leave out the risotto and serve the veal with Espagnole sauce mixed with cooked peas and chopped truffle.

No. 83. Bragiuole di Vitello (Veal Cutlets)

Ingredients: Veal, salt, pepper, butter, bacon, carrots, flour, Chablis, water, lemon.

Cut a bit of veal steak into pieces the size of small cutlets, salt and pepper them, and put them in a wide low stewpan. Add two ounces of butter, a cut-up carrot, and some bits of bacon also cut up. When they are browned, add a spoonful of flour, half a glass of Chablis, and half a glass of water, and cook on a slow fire for half an hour, then take out the cutlets, reduce the sauce, and pass it through a sieve. Put it back on the fire and add an ounce of butter and a good squeeze of lemon, and when hot pour it over the cutlets.

No. 84. Costolette alla Manza (Veal Cutlets)

Ingredients: Veal cutlets (fowl or turkey cutlets), forcemeat, truffles, mushrooms, tongue, parsley, pasta marinate (No. 17).

Cut a few horizontal lines along your cutlets, and on each put a little veal or fowl forcemeat, to which add in equal quantities chopped truffles, tongue, mushrooms, and a little parsley. Over this put a thin layer of pasta marinate, and fry the cutlets on a slow fire.

No. 85. Vitello alla Pellegrina (Breast of Veal)

Ingredients: Breast of veal, butter, onions, sugar, stock, red wine, mushrooms, bacon, salt, flour, bay leaf.

Roast a bit of breast of veal, then glaze over two Spanish onions with butter and a little sugar, and when they are a good colour pour a teacup of stock and a glass of Burgundy over them, and add a few mushrooms, a bay leaf, some salt, and a few bits of bacon. When the mushrooms and onions are cooked, skim off the fat and thicken the sauce with a little flour and butter fried together; pour it over the veal and put the onions and mushrooms round the dish.

No. 86. Frittura Piccata al Marsala (Fillet of Veal)

Ingredients: Veal, butter, Marsala, stock, lemon, bacon.

Cut a tender bit of veal steak into small fillets, cut off all the fat and stringy parts, flour them and fry them in butter. When they are slightly browned add a glass of Marsala and a teacup of good stock, and fry on a very hot fire, so that the fillets may remain tender. Take them off the fire, put a little roll of fried bacon on each, add a squeeze of lemon juice, and serve.

No. 87. Polpettine Distese (Veal Olives)

Ingredients: Veal steak, butter, bread, eggs, pistacchio nuts, spice, parsley.

Cut some slices of veal steak very thin as for veal olives, and spread them out in a well-buttered stewpan. On each slice of veal put half a spoonful of the following mixture: Pound some crumb of bread and mix it with a whole egg; add a little salt, some pistacchio nuts, herbs, and parsley chopped up, and a little butter. Roll up each slice of veal, cover with a sheet of buttered paper, put the cover on the stewpan and cook for three-quarters of an hour in two ounces of butter on a slow fire. Thicken the sauce with a dessertspoonful of flour and butter fried together.

No. 88. Coste di Vitello Imboracciate (Ribs of Veal)

Ingredients: Ribs of veal, butter, eggs, Parmesan, bread crumbs, parsley.

Cut all the sinews from a piece of neck or ribs of veal, cover the meat with plenty of butter and half cook it on a slow fire, then let it get cold. When cold, egg it over and roll it in bread crumbs mixed with a tablespoonful of grated Parmesan; fry in butter and serve with a garnish of fried parsley and a rich sauce. A dessert-spoonful of New Century sauce mixed with quarter of a pint of good thick stock makes a good sauce. (See No. 226.)

No. 89. Costolette di Montone alla Nizzarda (Mutton Cutlets)

Ingredients: Mutton cutlets, butter, olives, mushrooms, cucumbers.

Trim as many cutlets as you require, and marinate them in vinegar, herbs, and spice for two hours. Before cooking wipe them well and then saute them in clarified butter, and when they are well coloured on both sides and resist the pressure of the finger, drain off the butter and pour four tablespoonsful of Espagnole sauce (No. 1) with a teaspoonful of vinegar and six bruised pepper corns over them. Arrange them on a dish, putting between each cutlet a crouton of fried bread, and garnish with olives stuffed with chopped mushrooms and with slices of fried cucumber.

No. 90. Petto di Castrato all'Italiana (Breast of Mutton)

Ingredients: Breast of mutton, veal, forcemeat, eggs, herbs, spice, Parmesan.

Stuff a breast of mutton with veal forcemeat mixed with two eggs beaten up, herbs, a little spice, and a tablespoonful of grated Parmesan, braize it in stock with a bunch of herbs and two onions. Serve with Italian sauce (No. 6).

No. 91. Petto di Castrato alla Salsa piccante (Breast of Mutton)

Ingredients: Same as No. 90.

When the breast of mutton has been stuffed and cooked as above, let it get cold and then cut it into fillets, flour them over, fry in butter, and serve with tomato sauce piquante (No. 10), or one dessert-spoonful of New Century sauce in a quarter pint of good stock or gravy.

No. 92. Tenerumi d'Agnello alla Villeroy (Tendons of Lamb)

Ingredients: Tendons of lamb, eggs, bread crumbs, truffles, butter, stock, Villeroy sauce.

Slightly cook the tendons (the part of the breast near the ribs) of lamb, press them between two dishes till cold, then cut into a good shape and dip them into a Villeroy sauce (No. 18) egg and bread-crumb, and saute them in butter. When about to serve, put them in a dish with very good clear gravy. A teaspoonful of chopped mint and a tablespoonful of chopped truffles mixed with the bread crumbs will be a great improvement.

No. 93. Tenerumi d' Agnello alla Veneziana (Tendons of Lamb)

Ingredients: Tendons of lamb, butter, parsley, onions, stock.

Fry the tendons of lamb in butter together with a teaspoonful of chopped parsley and an onion. Serve with good gravy.

No. 94. Costolette d' Agnello alla Costanza (Lamb Cutlets)

Ingredients: Lamb cutlets, butter, stock, cocks' combs, fowl's liver, mushrooms.

Fry as many lamb cutlets as you require very sharply in butter, drain off the butter and replace it with some very good stock or gravy. Make a ragout of cocks' combs, bits of fowl's liver and mushrooms all cut up; add a white sauce with half a gill of cream mixed with it, and with this mask the cutlets, and saute them for fifteen minutes.

Tongue, Sweetbread, Calf's Head, Liver, Sucking Pig, &C.

No. 95. Timballo alla Romana

Ingredients: Cold fowl, game, or sweetbread, butter, lard, flour, Parmesan, truffles, macaroni, onions, cream.

Make a light paste of two ounces of butter, two of lard, and half a pound of flour, and put it in the larder for two hours. In the meantime boil a little macaroni and let it get cold, then line a plain mould with the paste, and fill it with bits of cut-up fowl, or game, or sweetbread, bits of truffle cut in small dice, grated Parmesan, and a little chopped onion. Put these ingredients in alternately, and after each layer add enough cream to moisten. Fill the mould quite full, then roll out a thin paste for the top and press it well together at the edges to keep the cream from boiling out. Bake it in a moderate oven for an hour and a half, turn it out of the mould, and serve with a rich brown sauce. Decorate the top with bits of red tongue and truffles cut into shapes or with a little chopped pistacchio nut.

No. 96. Timballo alla Lombarda

Ingredients: Macaroni, fowl or game, eggs, stock, Velute sauce (No. 2), tongue, butter, truffles.

Butter a smooth mould, then boil some macaroni, but take care that it is in long pieces. When cold, take the longest bits and line the bottom of the mould, making the macaroni go in circles; and when you come to the end of one piece, join on the next as closely as possible until the whole mould is lined; paint it over now and then with white of egg beaten up; then mask the whole inside with a thin layer of forcemeat of fowl, which should also be put on with white of egg to make it adhere; then cut up the bits of macaroni which remain, warm them up in some good fowl stock and Velute sauce much reduced, a little melted butter, some bits of truffle cut into dice, tongue, fowl, or game also cut up in pieces. When the mould is full, put on another layer of forcemeat, steam for an hour, then turn out and serve with a very good brown sauce.

No. 97. Lingua alla Visconti (Tongue)

Ingredients: Tongue, glaze, bread, spinach, white grapes, port.

Soak a smoked tongue in fresh water for forty-eight hours, then boil it till it is tender. Peel off the skin, cut the tongue in rather thick slices, and glaze them. Prepare an oval border of fried bread, cover it with spinach about two inches thick, and on this arrange the slices of tongue. Fill in the centre of the dish with white grapes cooked in port or muscat.

No. 98. Lingua di Manzo al Citriuoli (Tongue with Cucumber)

Ingredients: Ox tongue, salt, pepper, nutmeg, parsley, bacon, veal, carrots, onions, thyme, bay leaves, cloves, stock.

Gently boil an ox tongue until you can peel off the skin, then lard it, season it with salt, pepper, nutmeg, and chopped parsley, and boil it with some bits of bacon, ham, veal, a carrot, an onion, two bay leaves, thyme and two cloves. Pour some good stock over it and let it simmer gently until it is cooked. Put the tongue on a dish and garnish it with slices of fried cucumber. Boil the cucumber for five minutes before you fry it, to take away the bitter taste. Serve the tongue with a sauce piquante, made with one dessert-spoonful of New Century sauce to a quarter pint of good Espangole sauce (No. 1).

No. 99. Lingue di Castrato alla Cuciniera (Sheep's Tongues)

Ingredients: Sheep's tongues, bacon, beef, onions, herbs, spice, eggs, butter, flour.

Cook three or four sheep's tongues in good stock, and add some slices of bacon, bits of beef, two onions, a bunch of herbs, and a pinch of spice. Let them get cold, flour them and mask them with egg beaten up and fry quickly in butter. Serve with Italian sauce (No. 6)

No. 100. Lingue di Vitello all'Italiana (Calves' Tongues)

Ingredients: Calves' tongues, salt, butter, stock, water, glaze, potatoes, ham, truffles, sauce piquante.

Rub a good handful of salt into two or three calves' tongues and leave them for twenty-four hours, then wash off all the salt and soak them in fresh water for two hours. Stew them gently till tender, take them out, skin and braize them in butter and good stock for half an hour. Let them get cold and cut them into slices about half an inch thick; put the slices into a buttered saute-pan and cover them with a good thick glaze; let them get quite hot and then arrange them on a border of potatoes, and garnish each slice with round shapes of cooked ham and truffle. Fill the centre with any vegetables you like; fried cucumber is excellent, but if you use it do not forget to boil it for five minutes before you fry it to take away the bitter taste. Serve with a sauce piquante (No. 10, or No. 226).

No. 101. Porcelletto alla Corradino (Sucking Pig)

Ingredients: Sucking pig, ham, eggs, Parmesan, truffles, mushrooms, garlic, bay leaves, coriander seeds, pistacchio nuts, veal forcemeat, suet, bacon, herbs, spice.

Bone a sucking pig, remove all the inside and fill it with a stuffing made of veal forcemeat mixed with a little chopped suet, ham, bacon, herbs, two tablespoonsful of finely chopped pistacchio nuts, a pinch of spice, six coriander seeds, two tablespoonsful of grated Parmesan, cuttings of truffles and mushrooms all bound together with eggs. Sew the pig up and braize it in a big stewpan with bits of bacon, a clove of garlic with two cuts, a bunch of herbs and one bay leaf, for half an hour. Then pour off the gravy, cover the pig with well-buttered paper, and finish cooking it in the oven. Garnish the top with vegetables and truffles cut into shapes, slices of lemon and sprigs of parsley. Serve with a good sauce piquante (No. 229). Do not leave the garlic in for more than ten minutes.

No. 102. Porcelletto da Latte in Galantina (Sucking Pig)

Ingredients: Sucking pig, forcemeat of fowl, bacon, truffles, pistacchio nuts, ham, lemon, veal, bay leaves, salt, carrots, onions, shallots, parsley, stock, Chablis, gravy.

Bone a sucking pig all except its feet, but be careful not to cut the skin on its back. Lay it out on a napkin and line it inside with a forcemeat of fowl and veal about an inch thick, over this put a layer of bits of marinated bacon, slices of truffle, pistacchio nuts, cooked ham, and some of the flesh of the pig, then another layer of forcemeat until the pig's skin is fairly filled. Keep its shape by sewing it lightly together, then rub it all over with lemon juice and cover it with slices of fat bacon, roll it up and stitch it in a pudding cloth. Then put the bones and cuttings into a stewpan with bits of bacon and veal steak cut up, two bay leaves, salt, a carrot, an onion, a shallot, and a bunch of parsley. Into this put the pig with a bottle of white wine and sufficient stock to cover it, and cook on a slow fire for three hours. Then take it out, and when cold take off the pudding-cloth. Pass the liquor through a hair sieve, and, if necessary, add some stock; reduce and clarify it. Decorate the dish with this jelly and serve cold.

No. 103. Ateletti alla Sarda

Ingredients: Veal or fowl, ox palates, stock, tongue, truffles, butter, mushrooms, sweetbread.

Soak two ox palates in salted water for four hours, then boil them until the rough skin comes off, and cook them in good stock for six hours, press them between two plates and let them get cold. Roll some forcemeat of veal or fowl in flour, cut it into small pieces about the size of a cork, boil them in salted water, let them get cold and cut them into circular pieces. Cut the ox palates also into circular pieces the same size as the bits of forcemeat, then thinner circles of cooked tongue and truffles. String these pieces alternately on small silver skewers. Reduce to half its quantity a pint of Velute sauce (No. 2), and add the cuttings of the truffles, mushroom trimmings, bits of sweetbread, and a squeeze of lemon juice. Let it get cold and then mask the atelets (or skewers with the forcemeat, &c.) with it, and fry them quickly in butter. Fry a large oval crouton of bread, scoop out the centre and fill it with fried slices of cucumber and truffles boiled in a little Chablis. Stick the skewers into the crouton and pour the sauce round it.

For a maigre dish use fillets of fish, truffles, mushrooms, and Bechamel sauce (No. 3). The cucumber should be boiled for five minutes before it is fried.

No. 104. Ateletti alla Genovese

Ingredients: Veal, sweetbread, calf's brains, ox palates, mushrooms, fonds d'artichauds, cocks' combs, eggs, Parmesan, bread crumbs.

Cook two ox palates as in the last recipe, then take equal quantities of veal steak, sweetbread, calf's brains, equal quantities of mushrooms, fonds d'artichauds, and cocks' combs. Fry them all in butter except the palates, but be careful to put the veal in first, as it requires longer cooking; the brains should go in last. Then put all these ingredients on a cutting board and add the palates (cooked separately); cut them all into pieces of equal size, either round or square, but keep the ingredients separate, and string them alternately on silver skewers, as in the last recipe. Then pound up all the cuttings and add a little crumb of bread soaked in stock, the yolks of three eggs, the whites of two well beaten up, two dessert-spoonsful of grated Parmesan, salt to taste, and chopped truffles. Mix all this well together and mask the atelets with it; egg and bread crumb them and fry in butter. When they are a good colour, serve with fried parsley.

No. 105. Testa di Vitello alla Sorrentina (Calf's Head)

Ingredients: Calf's head, veal, sweetbread, truffles, mushrooms, pistacchio nuts, eggs, herbs, spice, stock, bacon, ham.

Boil a half calf's head well, and when it is half cold, bone it and fill it with a stuffing of veal, the calf's brains, sweetbread, truffles, mushrooms, pistacchio nuts, the yolks of two eggs, herbs, and a little spice. Then stitch it up and braize it in good stock, with some slices of bacon, ham, and a bunch of herbs. Serve with brain sauce mixed with cream.

No. 106. Testa di Vitello con Salsa Napoletana (Calf's Head)

Ingredients: Calf's head, calf's liver, bacon, suet, truffles, almonds, olives, calf's brains, capers, spice, coriander seeds, herbs, ham, stock.

Boil half a calf's head, bone it and fill it with a stuffing made of four ounces of calf's liver, well chopped up and pounded in a mortar; two ounces of bacon, one ounce of suet, three truffles, six almonds, three olives, six coriander seeds, six capers, the calf's brains, a pinch of spice and a teaspoonful of chopped herbs. Roll up the head, tie it up and put it into a stewpan with some bits of bacon, ham, and very good stock, and stew it slowly. Serve with Neapolitan sauce (No.12), or with tomato sauce piquante (No. 10).

No. 107. Testa di Vitello alla Pompadour (Calf's Head)

Ingredients: Calf's head, calf's brains, cream, eggs, truffles, cinnamon, stock, butter, Parmesan.

Boil and bone half a calf's head and fill it with a stuffing made of the calf's brains, a gill of cream, the yolks of two eggs, two truffles cut up, a little chopped ham, and a tiny pinch of cinnamon. Boil it in good stock, and when it is sufficiently cooked take it out and mask it all over with a mixture of butter, yolk of egg, and a tablespoonful of grated Parmesan, then brown it in the oven and serve hot.

No. 108. Testa di Vitello alla Sanseverino (Calf's Head)

Ingredients: Calf's head, sweetbread, fowl's liver, anchovies, herbs, capers, garlic, bacon, ham, Malmsey or Muscat.

Boil and bone half a calf's head, and fill it with a stuffing made of half a pound of sweetbread, a fowl's liver, two anchovies, a teaspoonful of chopped herbs, a few chopped capers, and the calf's brains. Roll the head up, stitch it together and braize it in half a tumbler of Malmsey or Australian Muscat (Burgoyne's), half a cup of very good white stock, some bits of ham and bacon, and a clove of garlic with two cuts. Cook it gently for four hours and serve it with its own sauce. Do not leave the garlic in longer than ten minutes.

No. 109. Testa di Vitello in Frittata (Calf's Head)

Ingredients: Calf's head, eggs, Parmesan, ham, pepper, butter, croutons.

A good rechauffe' of calf's head may be made in the following manner: After the head has been well boiled in good stock, cut it into slices and mask these with a mixture of eggs well beaten up, grated Parmesan, pepper, and chopped ham. Fry in butter, and garnish with fried parsley and fried croutons. Serve with a sauce made of a quarter of a pint of good Bechamel (No. 3) and a dessertspoonful of New Century sauce.

No. 110. Zampetti (Calves' Feet)

Ingredients: Calves' or pigs' feet, butter, leeks or small onions, parsley, salt, pepper, stock, tomatoes, eggs, cheese, cinnamon.

Blanch and bone two or more calves' or pigs' feet and put them into a stewpan with butter, leeks, or onions, chopped parsley, salt, pepper, and a little stock. Let them boil till the liquid is somewhat reduced, then add good meat gravy and two tablespoonsful of tomato puree, and just before taking the stewpan off the fire, add the yolks of two eggs beaten up, a tablespoonful of grated cheese, and a tiny pinch of cinnamon. Mix all well together and serve very hot.

No. 111. Bodini Marinati

Ingredients: Veal forcemeat, truffles, sweetbread, mushrooms, herbs, flour, pasta marinate (No. 17), tongue, butter.

Make a mixture of truffles, tongue, sweetbread, mushrooms, and herbs, all chopped up, and add it to a forcemeat of veal, the proportions being two-thirds veal forcemeat and the other ingredients one third. Mix this well and form it into little balls about the size of a pigeon's egg, flour them and mask them all over with pasta marinate (No. 17). Fry them in butter over a slow fire, so that the balls may be well cooked through, and when they are the right colour dry them in a napkin and serve very hot.

These bodini may be made with various ingredients; they will be most delicate with a forcemeat of fowl and bits of brain mixed with herbs, truffle, cooked ham, or tongue. They are also excellent made with fish (sole, mullet, turbot, &c.), either cooked or raw, and marinated in lemon, salt, pepper, oil, nutmeg, and parsley.

No. 112. Animelle alla Parmegiana (Sweetbread)

Ingredients: Sweetbread, bread crumbs, Parmesan, butter.

Blanch as many sweetbreads as you require, and then roll them in bread crumbs mixed with grated Parmesan, salt, and pepper; wrap them up in buttered grease-proof paper and grill them. When they are cooked, take off the paper, and serve with a good sauce in a sauce-boat.

No. 113. Animelle in Cartoccio (Sweetbread)

Ingredients: Sweetbread, butter, herbs, salt, pepper, bread crumbs, Parmesan, lemons, gravy, tomatoes.

Blanch a pound of sweetbread cuttings, mix it with two ounces of melted butter, chopped herbs, salt, and pepper, and put it into paper souffle cases. Then strew over each some bread crumbs mixed with grated Parmesan, put the cases in the oven, and when they are browned serve either with good gravy and lemon juice or with tomato sauce (No. 9).

No. 114. Animelle all'Italiana (Sweetbread)

Ingredients: Sweetbread, butter, onions, salt, herbs, eggs, glaze, Risotto (No. 190), truffles, quenelles of fowl, Espagnole sauce, white sauce.

Blanch as many sweetbreads as you require, cut them into quarters and saute them in butter with a small onion cut up, salt, and a bunch of herbs. Then pour over them two cups of white sauce and cook gently for twenty minutes; take out the sweetbreads and put them in a stewpan. Reduce the sauce, and add to it a mixture made of the yolks of four eggs, one and a half ounce of butter and a teaspoonful of glaze; pass it through a sieve, pour it over the sweetbreads, and keep them warm in a bain-marie. Have ready a good Risotto all'Italiana (No. 190), and put it into a border mould (but first decorate the inside of the mould with slices of truffle), put it in a moderate oven, and when it is warm turn it out on a dish. Place the sweetbreads on the risotto and fill in the centre with quenelles of fowl and Espagnole sauce (No. 1).

No. 115. Animelle Lardellate (Sweetbread)

Ingredients: Sweetbreads, larding, bacon, stock, a macedoine of vegetables.

Blanch two sweetbreads, lard them, and cook them very slowly in good stock. Skim the stock and reduce it to a glaze to cover the sweetbreads. Then cut them into three or four pieces and arrange them round a dish, but see that the larding is well glazed over. In the centre of the dish place a piece of bread in the shape of a cup and fill this with a macedoine of vegetables.

No. 116. Frittura di Bottoni e di Animelle (Sweetbread and Mushrooms)

Ingredients: Sweetbread, fresh button mushrooms, flour, bread crumbs, salt, pepper, parsley, butter, lemons.

Peel some button mushrooms and cut them in halves. Boil a sweetbread, and cut it into pieces about the same size as the mushrooms, flour, egg, and bread crumb them, and fry in butter; then serve with a garnish of fried parsley. Hand cut lemons with this dish.

No. 117. Cervello in Fili serbe (Calf's Brains)

Ingredients: Calf's brains, stock, butter, parsley, lemon.

Boil half a calf's brain in good stock for ten minutes then drain and pour a little melted butter and the juice of half a lemon over the brain; add some chopped parsley fried for one minute in butter, and serve as hot as possible.

No. 118. Cervello alla Milanese (Calf's Brains)

Ingredients: Calf s brains, eggs, bread crumbs, butter.

Scald a calf's brain and let it get cold. Wipe it on a cloth, and get it as dry as possible, then cut it into pieces about the size of a walnut, egg and bread crumb them, fry in butter, and strew a little salt over them.

No. 119. Cervello alla Villeroy (Calf's Brains)

Ingredients: Calf's brains, eggs, flour, mushrooms, Velute sauce.

Scald a calf's brain, and when cold cut it up and mask each piece with a thick sauce made of well-reduced Velute (No. 2), mixed with chopped cooked mushrooms; flour them over and dip them into the yolk of an egg, and fry as quickly as possible.

No. 120. Frittura of Liver and Brains

Ingredients: Calf's liver and brains (or lamb's or pig's fry), butter, ham, flour, puff pastry.

Cut up half a pound of liver in small slices, flour and fry them in butter or dripping, together with a calf's or pig's or sheep's brain, previously scalded and also cut up. Serve with bits of fried ham and little diamond-shaped pieces of puff pastry.

No. 121. Cervello in Frittata Montano (Calf's Brains)

Ingredients: Calf's brains, stock, cream, eggs, spice, Parmesan, butter.

Boil a calf's brain in good stock for ten minutes, let it get cold, cut it up into little balls, and mask each piece with a mixture made of half a gill of cream, the yolks of two eggs, a little spice, a tablespoonful of grated Parmesan, and the whites of two eggs well beaten up. Fry the balls in butter, and serve as hot as possible. You may mask and cook the calf's brain without cutting it up, if you prefer it so.

No. 122. Marinata di Cervello alla Villeroy (Calf's Brains)

Ingredients: Calf's brains, stock, Bechamel sauce, eggs, butter, lemon, forcemeat of fowl, flour.

Boil a calf's or sheep's brain in good stock, wipe it well, and cut it up. Reduce a pint of Bechamel (No. 3), and add to it the yolks of three eggs, an ounce of butter, and the juice of a lemon. When it boils throw in the cut-up brain; let it cool, then take out the brain and form it into little balls about the size of a small walnut. Make a forcemeat of fowl, and add a dessert-spoonful of flour to it, and spread it out very thin on a paste-board, and into this wrap the balls of brain, each separately. Dip them into a pasta marinate (No. 17), and fry them a golden brown.

No. 123. Minuta alla Milanese (Lamb's Sweetbread)

Ingredients: Lamb's sweetbread, butter, onions, stock, Chablis, salt, lemon, herbs, cocks' combs, fowls' livers.

Cut up equal quantities of lamb's sweetbreads, cocks' combs, fowls' livers in pieces about the size of a filbert, flour and fry them slightly in butter and a small bit of onion, add half a glass of Chablis, a cup of good stock, and a bunch of herbs. Reduce the sauce, and thicken it with a tablespoonful of butter and flour fried together. Make a border of Risotto all'Italiana (No. 190), and put the sweetbread, &c., together with the sauce in the centre.

No. 124. Animelle al Sapor di Targone (Lamb's Fry)

Ingredients: Lamb's fry, ham, garlic, larding bacon, spice, herbs, butter, flour, stock.

The lamb's fry should be nearly all sweetbread, and very little liver. Lard each piece with bacon and ham, and roll it in chopped herbs and a pinch of pounded spice. Then dip it in flour and braize in good stock, to which add three ounces of butter, some bits of bacon, ham, a bay leaf, herbs, and a clove of garlic with two cuts. Cook until the fry is well glazed over, and serve with Tarragon sauce (No. 8). Do not leave the garlic in longer than ten minutes.

No. 125. Fritto Misto alla Villeroy

Ingredients: Cocks' combs, calf's brains, sweetbread, stock, truffles, mushrooms, Villeroy, eggs, bread crumbs.

Cook some big cocks' combs, bits of calf s brains, and sweetbread in good stock, then drain them and marinate them slightly in lemon juice and herbs. Prepare a Villeroy (No. 18), and add to it cuttings of sweetbread, brains, truffles, mushrooms, &c. When it is cold, mask the cocks' combs and other ingredients with it, egg and bread-crumb them, and fry them a golden brown.

No. 126. Fritto Misto alla Piemontese

Ingredients: Sweetbread, calf s brains, ox palate, flour, eggs, Chablis, salt, herbs butter.

Make a thin paste with a tablespoonful of flour, the yolks of two eggs, two Spoonsful of Chablis, and a little salt. Mix this up well, and if it is too thick add a little water. Beat up the whites of the two eggs into a snow. In the meantime blanch a sweetbread, half a calf's brain, and a few bits of cooked ox palate; boil them all up with a bunch of herbs; cut them into pieces about the size of a walnut, and dip them into the paste so that each piece is well covered, then dip them into the beaten-up whites of egg, and fry them very quickly in butter. This fry is generally served with a garnish of French beans, which should not be cut up, but half boiled, then dried, floured over and fried together with the other ingredients. The ox palates should be boiled for at least six hours before you use them in this dish.

No. 127. Minuta di Fegatini (Ragout of Fowls' Livers)

Ingredients: Fowls' or turkeys' livers, flour, butter, parsley, onions, salt, pepper, stock, Chablis.

Cut the livers in half, flour them, and fry lightly in butter with chopped parsley, very little chopped onion, salt and pepper, then add a quarter pint of boiling stock and half a glass of Chablis, and cook until the sauce is somewhat reduced. You can also cook the livers simply in good meat gravy, but in this case they should not be floured. Serve with a border of macaroni (No. 183), or Risotto (No. 190), or Polenta (No. 187).

No. 128. Minuta alla Visconti (Chickens' Livers)

Ingredients: Fowls' livers, eggs, cheese, butter, cream, cayenne pepper.

Braize two fowls' livers in butter, then pound them up, and mix with a little cream, a tablespoonful of grated cheese and a dust of cayenne.

Spread this rather thickly over small squares of toast, and keep them hot whilst you make a custard with half an ounce of butter, an egg well beaten up, and a tablespoonful of cheese. Stir it over the fire till thick and then spread it on the hot toast. Serve very hot. This makes a good savoury.

No. 129. Croutons alla Principesca

Ingredients: Croutons, tongue, sweetbread, truffles, fowl or game, Velute sauce, stock, eggs, butter.

Fry a bit of bread in butter till it is a light brown colour, then cut it into heart-shaped pieces. Prepare a ragout with bits of tongue, sweetbread, fowl or game, truffles, two or three spoonsful of well-reduced Velute sauce (No. 2), and two or three of reduced gravy. Put a spoonful of the ragout in each crouton, and over it a layer of fowl forcemeat half an inch thick; trim the edges neatly, glaze them with the yolk of eggs beaten up, and put them in a buttered fire-proof dish in the oven for twenty minutes. Then glaze them with reduced stock and serve hot.

For a maigre dish use fish for the ragout and forcemeat.

No. 130. Croutons alla Romana

Ingredients: Bread, fowl forcemeat, tongue, truffles, herbs, cream, stock, butter, flour, eggs.

Cut a bit of crumb of bread into round or square shapes, and on each put a spoonful of fowl or rabbit forcemeat, a little chopped tongue, and a slight flavouring of chopped herbs; cover with a slice of bread the same shape as the underneath piece, put them in a buttered fireproof dish, and moisten them well with cream, butter, and stock. Cook until all the liquor is absorbed, but turn them over so that both sides may be well cooked, then flour and dip them into beaten-up eggs; fry them a good colour and serve very hot.

For a maigre dish use forcemeat of fish or lobster, and more cream instead of stock.

Fowl, Duck, Game, Hare, Rabbit, &c.

No. 131. Soffiato di Cappone (Fowl Souffle)

Ingredients: Fowl, Bechamel, stock, semolina flour, potatoes, salt, eggs, butter, smoked tongue or ham.

Prepare a puree of fowl or turkey and a small quantity of grated tongue or ham, and whilst you are pounding the meat add some good gravy or stock. Then make a Bechamel sauce (No. 3) and add two table-spoonsful of semolina flour, a boiled potato and salt to taste, boil it up and add the puree of fowl, then let it get nearly cold, add yolks of eggs and the white beaten up into a snow. (For one pint of the puree use the yolks of three eggs.) Pour the whole into a buttered souffle case, and half an hour before serving put it in a moderate oven and serve hot. You can use game instead of fowl, and serve in little souffle cases.

No. 132. Pollo alla Fiorentina (Chicken)

Ingredients: Fowl, butter, vegetables, rice or macaroni, peppercorns, stock, ham, tomatoes, bay leaves, onions, cloves, Liebig.

Roll up a fowl in buttered paper and put it in the oven in a fire-proof dish with all kinds of vegetables and a few peppercorns. Leave it there for about two hours, then put the fowl and vegetables into two quarts of good stock and let it simmer for one hour; serve on well-boiled rice or macaroni and pour the following sauce over it. Sauce: Two pounds tomatoes, one big cup of good stock, a quarter pound of chopped ham, three bay leaves, one onion stuck with cloves, one teaspoonful of Liebig. Simmer an hour and a half.

No. 133. Pollo all'Oliva (Chicken)

Ingredients: Fowl, onions, celery, salt, parsley, carrots, butter, stock, olives, tomatoes.

Cut up half an onion, a stick of celery, a sprig of parsley, a carrot, and cook them all in a quarter pound of butter. Into this put a fowl cut up and let it act brown all over, turn when necessary and then baste it with boiling stock. Add four Spanish olives cut up and four others pounded in a mortar, eight whole olives and three tablespoonsful of tomato puree reduced, and when the fowl is well cooked pour the sauce over it.

No. 134. Pollo alla Villereccia (Chicken)

Ingredients: Fowl, butter, flour, stock, bacon, ham, mushrooms, onions, cloves, eggs, cream, lemons.

Cut up a fowl into quarters and put it into a saucepan with three ounces of butter and a tablespoonful of flour Put it on the fire, and when it is well browned add half a pint of stock, bits of bacon and ham, butter, three mushrooms (previously boiled), an onion stuck with three cloves. When this is cooked skim off the grease, pass the sauce through a sieve, and add the yolks of two eggs mixed with two tablespoonsful of cream. Lastly, add a squeeze of lemon juice to the sauce and pour it over the fowl.

No. 135. Pollo alla Cacciatora (Chicken)

Ingredients: The same as No. 134 and tomatoes.

Cook the fowl exactly as above, but add either a puree of tomatoes or tomato sauce.

No. 136. Pollastro alla Lorenese (Fowl)

Ingredients: Fowl, butter, parsley, lemon, small onions, bread crumbs.

Cut up a fowl and put it into a frying pan with two ounces of butter, one onion cut up and a sprig of chopped parsley, salt and pepper; put it on the fire and cook it, but turn the pieces several times: then take them out and roll them whilst hot in bread crumbs, and fry them. Serve with cut lemons.

No. 137. Pollastro in Fricassea al Burro (Fowl)

Ingredients: Fowl, butter, fat bacon, ham, mushrooms, truffles, herbs, spice, gravy.

Cut up a fowl and cook it in a fricassee of butter, bacon, ham, herbs, mushrooms, truffles, spice, and good gravy or stock. Serve in its own gravy.

No. 138. Pollastro in istufa di Pomidoro (Braized Fowl)

Ingredients: Fowl, bacon, ham, bay leaf, spice, garlic, Burgundy, tomatoes.

Braize a fowl with bits of fat bacon, ham, a bay leaf, a clove of garlic with one cut in it, a pinch of spice, and a glass of Burgundy. Only leave the garlic in for five minutes. When cooked serve with tomato sauce (No. 9).

No. 139. Cappone con Riso (Capon with Rice)

Ingredients: Capon, veal forcemeat, fat bacon, stock, rice, truffles, mushrooms, cocks' combs, kidneys or fowls' liver, supreme sauce, milk, Chablis.

Stuff a fine capon with a good firm forcemeat made of veal, tongue, ham, and chopped truffles; cover it with larding bacon; tie it up in buttered paper, and cook it in very good white stock. In the meantime boil four ounces of rice in milk till quite stiff, mix in some chopped truffles, and make ten little timbales of it. Take out the capon when it is sufficiently cooked and place it on a dish; garnish it with cooked mushrooms, cocks' combs, kidneys, or fowls' livers, and pour a sauce supreme (No. 16) over it; round the dish place the timbales of rice, and between each put a whole truffle cooked in white wine. Serve a sauce supreme in a sauce bowl.

No. 140. Dindo Arrosto alla Milanese (Roast Turkey)

Ingredients: Turkey, sausage meat, prunes, chestnuts, a pear, butter, Marsala, salt, rosemary, bacon, carrot, onion, turnip, garlic.

Blanch for seven or eight minutes three prunes, quarter of a pound of sausage meat, three tablespoonsful of chestnut puree, two small slices of bacon, half a cooked pear, and saute them in butter; chop up the liver and gizzard of the turkey, mix them with the other ingredients, and add half a glass of Marsala; use this as a stuffing for the turkey, and first braize it for three quarters of an hour with salt, butter, a blade of rosemary, bits of fat bacon, a carrot, a turnip, an onion, three cloves, and a clove of garlic with a cut; then roast it before a clear fire for about twenty minutes; put it back into the sauce till it is ready to serve. Only leave the garlic in ten minutes.

No. 141. Tacchinotto all'Istrione (Turkey Poult)

Ingredients: A turkey poult, ham, mace, bay leaves, lemons, water, salt, onions, parsley, celery, carrots, Chablis.

Truss a turkey poult, and cover it all over with slices of ham or bacon, put two bay leaves and four slices of lemon on it, and sprinkle with a small pinch of mace, then sew it up tight in a dishcloth, and stew it in good stock, salt, an onion, parsley, a stick of celery, a carrot, and a pint of Chablis; cook for an hour, take it out of the cloth, and pour a good rich sauce over it. It is also good cold with aspic jelly.

No. 142. Fagiano alla Napoletana (Pheasant)

Ingredients: Pheasant, macaroni, gravy, butter, Parmesan, tomatoes.

Lard a pheasant, roast it, and serve it on a layer of macaroni cooked with good reduced gravy, two ounces of butter, a tablespoonful of grated Parmesan, and a puree of tomatoes. Serve with Neapolitan sauce (No. 12) in a sauce bowl.

No. 143. Fagiano alla Perigo (Pheasant)

Ingredients: Pheasant, butter, truffles, larding bacon, Madeira.

Make a mixture of three tablespoonsful of chopped truffles, three ounces of butter and a little salt, and with this stuff a pheasant. Then cover it with slices of fat bacon and keep it in a cool place till next day. A few hours before serving, roast the pheasant and baste it well with melted butter and a wine-glass of Madeira or Marsala. Make a crouton of fried bread the shape of your dish, and over this put a Layer of forcemeat of fowl and a number of small fowl quenelles; cover them with buttered paper, then put the dish in the oven for a few minutes so as to settle the forcemeat. When the pheasant is cooked, place it on the crouton and garnish it with slices of truffle which have been previously cooked in Madeira, and serve with a Perigord sauce.

No. 144. Anitra Selvatica (Wild Duck)

Ingredients: Wild duck, butter, fowls' livers, Marsala, gravy, turnips, carrots, parsley, mushrooms.

Cut a wild duck into quarters and put it into a stewpan with two fowls' livers cut up and fried in butter. When the pieces of duck are coloured on both sides, pour off the butter, and in its place pour a glass of Marsala, a cup of stock, and a cup of Espagnole sauce (No.1), and cook gently for ten minutes. In the meantime shape and blanch six young turnips and as many young carrots, put them into a stewpan, and on the top of them put the pieces of wild duck, liver, &c. Pass the liquor through a sieve and pour it over the wild duck, add a bunch of parsley and other herbs and five little mushrooms cut up, and cook on a slow fire for half an hour. Skim the sauce, pass it through a sieve and add a pinch of sugar. Put the pieces of wild duck in an entree dish, add the vegetables, &c., pour the sauce over and serve.

No. 145. Perniciotti alla Gastalda (Partridges)

Ingredients: Partridges, cauliflower, bacon, sausage, fowls' livers, carrots, onions herbs, stock, gravy, butter, Madeira.

Cut a cauliflower into quarters, blanch for a few minutes, drain, and put it into a saucepan with some bits of bacon. Let it drain on paper till dry, then arrange the bits in a circle in a deep stewpan, and in the centre put a small bit of sausage, the livers of the partridges, a fowl's liver cut up, a carrot, an onion, and a bunch of herbs. Cover about three-quarters high with good stock and gravy, put butter on the top and boil gently for an hour; then take out the sausage, replace it by two or three partridges, and simmer for three-quarters of an hour. In the meantime cut a sausage in thin slices and line a mould with it. When the birds are cooked, take them out, drain and cut them up, and fill the mould with alternate layers of partridge and cauliflower, and steam for half an hour. Five minutes before serving turn the mould over on a plate, but do not take it off, so as to let all the grease drain off. Cut up the fowls' and partridges' livers, make them into scallops and glaze them. Wipe off all the grease round the mould; take it off, garnish the dish with the scallops of liver and serve hot with an Espagnole sauce (No. 1) reduced, and add a glass of Madeira or Marsala, and a glass of essence of game to it. This is an excellent way of cooking an old partridge or pheasant.

No. 146. Beccaccini alla Diplomatica (Snipe)

Ingredients: Snipe, ham, larding bacon, herbs, Marsala, croutons, truffles, cocks' combs, mushrooms, sweetbread, tongue.

Truss fourteen snipe and cook them in a mirepoix made with plenty of ham, fat bacon, herbs, and a wine glass of Marsala. When they are cooked pour off the sauce, skim off the grease and reduce it. Take the two smallest snipe and make a forcemeat of them by pounding them in a mortar with the livers of all the snipe, then dilute this with reduced Espagnole sauce (No. 1) and add it to the first sauce. Cut twelve croutons of bread just large enough to hold a snipe each, and fry them in butter. Add some chopped herbs and truffles to the forcemeat, spread it on the croutons, and on each place a snipe and cover it with a bit of fat bacon and buttered paper. Put them in a moderate oven for a few minutes, arrange them on a dish, and pour some of their own sauce over them. Garnish the spaces between the croutons with white cocks' combs, mushrooms, and truffles. The truffles should be scooped out and filled with a little stuffing of sweetbread, tongue, and truffles mixed with a little of the sauce of the snipe. Serve the rest of the sauce in a sauce-boat.

No. 147. Piccioni alla minute (Pigeons)

Ingredients: Pigeons, butter, truffles, herbs, fowls' livers, sweetbread, salt, flour, stock, Burgundy.

Prepare two pigeons and put them into a stewpan with two ounces of butter, two truffles cut up, two fowls' livers, half-pound of sweetbread cuttings (boiled), a bunch of herbs and salt. Let them brown a little, then add a dessert-spoonful of flour mixed with stock, and half a glass of Burgundy, and stew gently for half an hour.

No. 148. Piccioni in Ripieno (Stuffed Pigeons)

Ingredients: Pigeons, sweetbread, parsley, onions, carrots, salt, pepper, bacon, stock, Chablis, fowls' livers, and gizzards.

Cut up a sweetbread, a fowl's liver and gizzard, an onion, a sprig of parsley, and add salt and pepper. Put this stuffing into two pigeons, tie larding bacon over them, and put them into a stewpan with a glass of Chablis, a cup of stock, an onion, and a carrot. When cooked pass the sauce through a sieve, skim it, add a little more sauce, and pour it over the pigeons.

No. 149. Lepre in istufato (Stewed Hare)

Ingredients: Hare, butter, onions, garlic, marjoram, celery, ham, salt, Chablis, stock, mushrooms, spice, tomatoes.

Put into a stewpan three ounces of butter, an onion cut up, a clove of garlic with a cut across it, a sprig of marjoram, and a little cut-up ham. Fry these slightly, put the hare cut up into the same stewpan, and let it get brown. Then pour a glass of Chablis and a glass of stock over it; add a little tomato sauce or a mashed-up tomato, a pinch of spice, and a few mushrooms; take out the garlic and let the rest stew gently for an hour or more. Keep the cover on the stewpan, but stir the stew occasionally.

No. 150. Lepre Agro-dolce (Hare)

Ingredients: Hare, vinegar butter, onion, ham, stock salt, sugar, chocolate, almonds, raisins.

Cut up a hare and wash the pieces in vinegar, then cook them in butter, chopped onion, some bits of ham stock and a little salt. Half fill a wine-glass with sugar and add vinegar until the glass is three-quarters full mix the vinegar and sugar well together, and when the hare is browned all over and nearly cooked, pour the vinegar over it and add a dessert spoonful of grated chocolate a few shredded almonds and stoned raisins. Mix all well together and cook for a few minutes more. This is a favourite Roman dish.

No. 151. Coniglio alla Provenzale (Rabbit)

Ingredients: Rabbit, flour butter, stock, Chablis, parsley onion, spice, mushrooms.

Cut up a rabbit, wipe the pieces, flour them over, and fry them in butter until they are coloured all over. Then pour a glass of Chablis over them, add some chopped parsley, half an onion, three mushrooms, salt, and a cup of good stock. Cover the stewpan and cook on a moderate fire for about three-quarters of an hour. Should the stew act too dry, add a spoonful of stock occasionally.

No. 152. Coniglio arrostito alla Corradino (Roast Rabbit)

Ingredients: Rabbit, pig's fry, butter, salt, pepper, fennel, bay leaf, onions.

Make a stuffing of pig's fry (previously cooked in butter), salt, pepper, fennel, an onion, all chopped up, and a bay leaf. With this stuff a rabbit well and braize it for half an hour, then roast it before a brisk fire and baste it well with good gravy. If you like, put in a clove of garlic with one cut whilst it is being braized, but only leave it in for five minutes. Serve with ham sauce (Salsa di prosciutto, No. 7.) A fowl may be cooked in this way.

No. 153. Coniglio in salsa Piccante (Rabbit)

Ingredients: Rabbit, butter, flour, celery, parsley, onion, carrot, mushrooms, cloves, spices, Burgundy, stock, capers, anchovies.

Cut up a rabbit, wipe the pieces well on a dishcloth, flour them over and put them into a frying-pan with two ounces of butter and fry for about ten minutes. Then add half a stick of celery, parsley, an onion, half a carrot, and three mushrooms, all cut up, three cloves, a pinch of spice and salt, a glass of Burgundy, and the same quantity of stock; cover the stewpan and cook for half an hour, then put the pieces of rabbit into another stewpan and pass the liquor through a sieve; press it well with a wooden spoon, so as to get as much through as possible, pour this over the rabbit and add four capers and an anchovy in brine pounded in a mortar, mix all well together, let it simmer for a few minutes, then serve hot with a garnish of croutons fried in butter.

Vegetables

No. 154. Asparagi alla salsa Suprema (Asparagus)

Ingredients: Asparagus, butter, nutmeg, salt, supreme sauce (No. 16) gravy, lemon, Parmesan.

Cut some asparagus into pieces about an inch long and cook them in boiling water with salt, then drain and put them into a saute pan with one and a half ounce of melted butter and sautez for a few minutes, but first add salt, a pinch of nutmeg, and a dust of grated cheese. Pour a little supreme sauce over them, and at the last add a little gravy, one ounce of fresh butter, and a squeeze of lemon juice.

No. 155. Cavoli di Bruxelles alla Savoiarda (Brussels Sprouts)

Ingredients: Brussels sprouts, butter, pepper, stock, Bechamel sauce, Parmesan, croutons.

Take off the outside leaves of half a pound of Brussels sprouts, wash and boil them in salted water. Let them get cool, drain, and put them in a pie-dish with two ounces of fresh butter, a quarter pint of very good stock, a little pepper, and a dust of grated Parmesan. When they are well glazed over, pour off the sauce, season with three tablespoonsful of boiling Bechamel sauce (No. 3), and serve with croutons fried in butter.

No. 156. Barbabietola alla Parmigiana (Beetroot)

Ingredients: Beetroot, white sauce, Parmesan, Cheddar.

Boil a beetroot till it is quite tender, peel it, cut into slices, put it in a fireproof dish, and cover it with a thick white sauce. Strew a little grated Parmesan and Cheddar over it. Put it in the oven for a few minutes, and serve very hot in the dish.

No. 157. Fave alla Savoiarda (Beans)

Ingredients: Beans, stock, a bunch of herbs, Bechamel sauce.

Boil one pound of broad beans in salt and water, skin and cook them in a saucepan with a quarter pint of reduced stock and a hunch of herbs. Drain them, take out the herbs, and season with two glasses of Bechamel sauce (No. 3).

No. 158. Verze alla Capuccina (Cabbage)

Ingredients: Cabbage or greens, anchovies, salt, butter, parsley, gravy, Parmesan.

Boil two cabbages in a good deal of water, and cut them into quarters. Fry two anchovies slightly in butter and chopped parsley, add the cabbages, and at the last three tablespoonsful of good gravy, two tablespoonsful of grated Parmesan, salt and pepper, and when cooked, serve.

No. 159. Cavoli fiodi alla Lionese (Cauliflower)

Ingredients: Cauliflower, butter, onions, parsley, lemon, Espagnole sauce.

Blanch a cauliflower and boil it, but not too much. Cut up a small onion, fry it slightly in butter and chopped parsley, and when it is well coloured, add the cauliflower and finish cooking it, then take it out, put it in a dish, pour a good Espagnole sauce (No. 1) over it, and add a squeeze of lemon juice.

No. 160. Cavoli fiodi fritti (Cauliflower)

Ingredients: Cauliflower or broccoli, gravy, lemon, salt, eggs, butter.

Break up a broccoli or cauliflower into little bunches, blanch them, and put them on the fire in a saucepan with good gravy for a few minutes, then marinate them with lemon juice and salt, let them get cold, egg them over, and fry in butter.

No. 161. Cauliflower alla Parmigiana

Ingredients: Cauliflower, butter, Parmesan, Cheddar, Espagnole, stock.

Boil a cauliflower in salted water, then sautez it in butter, but be careful not to cook it too much. Take it off the fire and strew grated Parmesan and Cheddar over it then put in a fireproof dish and add a good spoonful of stock and one of Espagnole (No. 1), and put it in the oven for ten minutes.

No. 162. Cavoli Fiori Ripieni

Ingredients: Cauliflower, butter, stock, forcemeat of fowl, tongue, truffles, mushrooms, parsley, Espagnole, eggs.

Break up a cauliflower into separate little bunches, blanch them, and put them in butter, and a quarter pint of reduced stock. Make a forcemeat of fowl, add bits of tongue, truffles, mushrooms, and parsley, all cut up small and mixed with butter. With this mask the pieces of cauliflower, egg and breadcrumb them, fry like croquettes, and serve with a good Espagnole sauce (No. 1).

No. 163. Sedani alla Parmigiana (Celery)

Ingredients: Celery, stock, ham, salt, pepper, Cheddar, Parmesan, butter, gravy.

Cut all the green off a head of celery, trim the rest. Cut it into pieces about four inches long, blanch and braize them in good stock, ham, salt, and pepper. When cooked, drain and arrange them on a dish, sprinkle with grated Parmesan and Cheddar, and add one and a half ounce of butter, then put them in the oven till they have taken a good colour, pour a little good gravy over them and serve.

No. 164. Sedani fritti all'Italiana (Celery)

Ingredients: Same as No. 163, eggs, bread crumbs, tomatoes.

Prepare a head of celery as above, and cut it up into equal pieces. Blanch and braize as above, and when cold egg and breadcrumb and sautez in butter. Serve with tomato sauce.

No. 165. Cetriuoli alla Parmigiana (Cucumber)

Ingredients: Cucumber, butter, cheese, gravy, salt, cayenne.

Cut a cucumber into slices about half an inch thick, boil for five minutes in salted water, drain in a sieve, and fry slightly in melted butter, then strew a little grated Parmesan over it, and add a good thick gravy, put it into the oven for ten minutes to brown, and serve as hot as possible.

No. 166. Cetriuoli alla Borghese (Cucumber)

Ingredients: Cucumber, cream, salt, Bechamel sauce, butter, Parmesan, cayenne pepper.

Cook a cucumber as in No. 165, braize it for five minutes, add to it a good rich Bechamel (No. 3), mixed with cream and grated Parmesan Spread this well over the cucumber, and put it into the oven for ten minutes keeping the rounds of cucumber separate, so as to arrange them in a circle on a very hot dish. Care should be taken not to cook the cucumber too long, or it will break in pieces and spoil the look of the dish.

No. 167. Carote al sughillo (Carrots)

Ingredients: Carrots, stock, butter, sausage, pepper.

Boil some young carrots in stock, slice them up, and put them in a stewpan with a sausage cut up; cook for quarter of an hour on a slow fire, then stir up the fire, and when the carrots and sausage are a good colour add a good Espagnole sauce (No. 1), and serve.

No. 168. Carote e piselli alla panna (Carrots and Peas)

Ingredients: Young carrots, peas, cream, salt.

Half cook equal quantities of peas and young carrots (the carrots should be cut in dice, and will require a little longer cooking), then put them together in a stewpan with three or four tablespoonsful of cream, and cook till quite tender. Serve hot.

No. 169. Verze alla Certosine (Cabbage)

Ingredients: Cabbage, butter, salt, leeks or shallots, sardines, cheese.

Any vegetable may be cooked in the following simple manner: Boil them well, then slightly fry a little bit of leek or shallot and a sardine in butter; drain the vegetables, put them in the butter, and cook gently so that they may absorb all the flavour, and at the last add a dust of grated cheese and a tiny pinch of spice.

No. 170. Lattughe al sugo (Lettuce)

Ingredients: Lettuce, Parmesan, bacon, stock, butter, croutons of bread, gravy.

Take off the outside leaves of a lettuce, blanch and drain them well. Put on each leaf a mixture of grated Parmesan, salt, little bits of chopped bacon or ham, add a little good stock, cover over with buttered paper, and cook in a hot oven for five minutes. Then drain off the stock and roll up each leaf with the bacon, &c., put them on croutons of fried bread and pour some good thick gravy over them.

No. 171 Lattughe farcite alla Genovese (Lettuce)

Ingredients: Lettuce, forcemeat of fowl or veal, ham, Espagnole sauce.

Prepare a lettuce as above, and spread on each leaf a spoonful of forcemeat of fowl or veal, add a little cooked ham chopped up, roll up the leaves, and cook as above. Drain them on a cloth, arrange them neatly on a dish, and pour some good Espagnole sauce (No. 1) over them.

No. 172. Funghi cappelle infarcite (Stuffed Mushrooms)

Ingredients: Mushrooms, bread, stock, garlic, parsley, salt, Parmesan, butter, eggs, cream.

Choose a dozen good fresh mushrooms, take off the stalks and put the tops into a saucepan with a little butter. See that they lie bottom upwards. Then cut up and mix together half the stalks of the mushrooms, a little bread crumb soaked in gravy, the merest scrap of garlic and a little chopped parsley. Put this into a separate saucepan and add to it two eggs, half a gill of cream, salt, and two tablespoonsful of grated Parmesan. Mix well so as to get a smooth paste and fill in the cavities of the mushrooms with it. Then add a little more butter, strew some bread crumbs over each mushroom, and cook in the oven for ten to fifteen minutes.

No. 173. Verdure miste (Macedoine of Vegetables)

Ingredients: Cauliflower, carrots, celery, spinach, butter, cream, pepper, Parmesan.

Boil some carrots, cauliflower, spinach, and celery (all cut up) in water. Then put them in layers in a buttered china mould, and between each layer add a little cream, pepper, and a little grated Parmesan and Cheddar. Fill the mould in this manner, and put it in the oven for half an hour, so that the vegetables may cook without adhering to the mould. Turn out and serve.

No. 174. Patate alla crema (Potatoes in cream)

Ingredients: Potatoes, butter, Parmesan, white stock, cream, pepper, salt.

Boil two pounds of potatoes in salted water for a quarter of an hour, peel and cut them into slices about the size of a penny, then arrange them in layers in a very deep fireproof dish (with a lid), and on each layer pour a little melted butter, a little good white stock and a dust of grated Parmesan. Reduce a pint and a half of cream to half its quantity, add a little pepper, and pour it over the potatoes. Put the dish in the oven for twenty minutes. Serve as hot as possible.

No. 175. Cestelline di patate alla giardiniera (Potatoes)

Ingredients: Potatoes, white stock, salt, butter, peas, asparagus, sprouts, beans, &c.

Choose some big sound potatoes, cut them in half and scoop out a little of the centre so as to form a cavity, blanch them in salted water and cook for a quarter of an hour in good white stock and a little butter. Then fill in the cavities with a macedoine of cooked vegetables and add a little cream to each.

No. 176. Patate al Pomidoro (Potatoes with Tomato Sauce)

Ingredients: Potatoes, butter, salt, tomatoes, lemon, stock.

Peel three or four raw potatoes, cut them in slices about the size of a five-shilling piece, then put them into a stewpan with two ounces of melted butter, and cook them gently until they are a good colour, add salt, drain off the butter, then glaze them by adding half a glass of good stock. Arrange them on a dish, pour some good tomato sauce over them, and add a little butter and a squeeze of lemon juice.

No. 177. Spinaci alla Milanese (Spinach)

Ingredients: Spinach, butter, Velute sauce, salt, pepper, flour, stock.

Wash three pounds of spinach at least six times, boil it in a pint of water, then mince it up very fine, pass it through a hair-sieve, and put it in a saucepan with one and a half ounces of butter, add a cupful of reduced Velute sauce (No. 2) with cream, salt, and pepper, add a dessert-spoonful of flour and butter mixed, and boil until the spinach is firm enough to make into a shape, garnish with hard-boiled eggs cut into quarters, and pour a good Espagnole sauce (No. 1) round the dish.

No. 178. Insalata di patate (Potato salad)

Ingredients: New potatoes, oil, white vinegar, onions, parsley, tarragon, chervil, celery, cream, salt, pepper, tarragon vinegar, watercress, cucumber, truffles.

Steam as many new potatoes as you require until they are well cooked, let them get cold, cut them into slices and pour three teaspoonsful of salad oil and one of white vinegar over them. Then rub a salad bowl with onion, put in a layer of the potato slices, and sprinkle with chopped parsley, tarragon, chervil, and celery, then another layer of potatoes until you have used all the potatoes; cover them with whipped cream seasoned with salt, pepper, and a little tarragon vinegar, and garnish the top with watercress, a few thin slices of truffle cooked in white wine, and some slices of cooked cucumber.

No. 179. Insalata alla Navarino (Salad)

Ingredients: Peas, bean onions, potatoes, tarragon, chives, parsley, tomatoes, anchovies, oil, vinegar, ham.

Mix a tablespoonful of chopped parsley, a teaspoonful of chopped onion, a teaspoonful of tarragon and chopped chives with half a gill of oil and half a gill of vinegar. Put this into a salad bowl with all sorts of cooked vegetables: peas, haricot beans, small onions, and potatoes cut up, and mix them w ell but gently, so as not to break the vegetables. Then add two or three anchovies in oil, and on the top place three or four ripe tomatoes cut in slices. A little cooked smoked ham cut in dice added to this salad is a great improvement.

No. 180. Insalata di pomidoro (Tomato Salad)

Ingredients: Tomatoes, mayonnaise, shallot, horseradish, gherkin, anchovies, fish, cucumber, lettuce, chervil, tarragon, eggs.

Mix the following ingredients: two anchovies in oil boned and minced, a gill of mayonnaise sauce, a little grated horseradish, very little chopped shallot, a little cold salmon or trout, and a small gherkin chopped. With this mixture stuff some ripe tomatoes. Then make a good salad of endive or lettuce, a teaspoonful of chopped tarragon and chervil, season it with oil, vinegar, salt, and pepper (the proportions should be three of oil to one of vinegar), put a layer of slices of cucumber in the salad, place the tomatoes on the top of these, and decorate them with hard-boiled eggs passed through a wire sieve.

No. 181. Tartufi alla Dino (Truffles)

Ingredients: Truffles, fowl forcemeat, champagne.

Allow one truffle for each person, scoop out the inside, chop it up fine and mix with a good forcemeat of fowl. With this fill up the truffles, place a thin layer of truffle on the top of each, and cook them in champagne in a stewpan for about half an hour. Then take them out, make a rich sauce, to which add the champagne you have used and some of the chopped truffle, put the truffles in this sauce and keep hot for ten minutes. Serve in paper souffle cases.

Macaroni, Rice, Polenta, and Other Italian Pastes[*]

> * Italian pastes of the best quality can be obtained at Cosenza's, Wigmore Street, NW. For the following dishes, tagliarelle and spaghetti are recommended.

No. 182. Macaroni with Tomatoes

Ingredients: Macaroni, tomatoes, butter, onion, basil, pepper, salt.

Fry half an onion slightly in butter, and as soon as it is coloured add a puree of two big cooked tomatoes. Then boil quarter of a pound of macaroni separately, drain it and put it in a deep fireproof dish, add the tomato puree and three tablespoonsful of grated Parmesan and Cheddar mixed, and cook gently for a quarter of an hour before serving. This dish may be made with vermicelli, spaghetti, or any other Italian paste.

No. 183. Macaroni alla Casalinga

Ingredients: Macaroni, butter, stock, cheese, water, salt, nutmeg.

Cut up a quarter pound of macaroni in small pieces and put it in boiling salted water. When sufficiently cooked, drain and put it into a saucepan with two ounces of butter, add good gravy or stock, three tablespoonsful of grated Parmesan and Cheddar mixed, and a tiny pinch of nutmeg. Stir over a brisk fire, and serve very hot.

No. 184. Macaroni al Sughillo

Ingredients: Macaroni, stock, tomatoes, sausage, cheese.

Half cook four ounces of macaroni, drain it and put it in layers in a fireproof dish, and gradually add good beef gravy, four tablespoonsful of tomato puree, and thin slices of sausage. Sprinkle with grated Parmesan and Cheddar, and cook for about twenty minutes. Before serving pass the salamander over the top to brown the macaroni.

No. 185. Macaroni alla Livornese

Ingredients: Macaroni, mushrooms, tomatoes, Parmesan, butter, pepper, salt, milk.

Boil about four ounces of macaroni, and stew four or five mushrooms in milk with pepper and salt. Put a layer of the macaroni in a buttered fireproof dish, then a layer of tomato puree, then a layer of the mushrooms and another layer of macaroni. Dust it all over with grated Parmesan and Cheddar, put it in the oven for half an hour, and serve very hot.

No. 186. Tagliarelle and Lobster

Ingredients: Tagliarelle, lobster, cheese, butter.

Boil half a pound of tagliarelle, and cut up a quarter of a pound of lobster. Butter a fireproof dish, and strew it well with grated Parmesan and Cheddar mixed, then put in the tagliarelle and lobster in layers, and between each layer add a little butter. Strew grated cheese over the top, put it in the oven for twenty minutes, and brown the top with a salamander.

No. 187. Polenta

Polenta is made of ground Indian-corn, and may be used either as a separate dish or as a garnish for roast meat, pigeons, fowl, &c. It is made like porridge; gradually drop the meal with one hand into boiling stock or water, and stir continually with a wooden spoon with the other hand. In about a quarter of an hour it will be quite thick and smooth, then add a little butter and grated Parmesan, and one egg beaten up. Let it get cold, then put it in layers in a baking-dish, add a little butter to each layer, sprinkle with plenty of Parmesan, and bake it for about an hour in a slow oven. Serve hot.

No. 188. Polenta Pasticciata

Ingredients: Polenta, butter, cheese, mushrooms, tomatoes.

Prepare a good polenta as above, put it in layers in a fireproof dish, and add by degrees one and a half ounces of melted butter, two cooked mushrooms cut up, and two tablespoonsful of grated cheese. (If you like, you may add a good-sized tomato mashed up.) Put the dish in the oven, and before serving brown it over with salamander.

No. 189. Battuffoli

Ingredients: Polenta, onion, butter, salt, stock, Parmesan.

Make a somewhat firm polenta (No. 187) with half a pound of ground maize and a pint and a half of salted water, add a small onion cut up and fried in butter, and stir the polenta until it is sufficiently cooked. Then take it off the fire and arrange it by spoonsful in a large fireproof dish, and give each spoonful the shape and size of an egg. Place them one against the other, and when the first layer is done, pour over it some very good gravy or stock, and plenty of grated Parmesan. Arrange it thus layer by layer. Put it into the oven for twenty minutes, and serve very hot.

No. 190. Risotto all'Italiana

Ingredients: Rice, an onion, butter, stock, tomatoes, cheese.

Fry a small onion slightly in butter, then add half a pint of very good stock. Boil four ounces of rice, but do not let it get pulpy, add it to the above with three medium-sized tomatoes in a puree. Mix it all up well, add more stock, and two tablespoonsful of grated Parmesan and Cheddar mixed, and serve hot.

No. 191. Risotto alla Genovese

Ingredients: Rice, beef or veal, onions, parsley, butter, stock, Parmesan, sweetbread or sheep's brains.

Cut up a small onion and fry it slightly in butter with some chopped parsley, add to this a little veal, also chopped up, and a little suet. Cook for ten minutes and then add two ounces of rice to it. Mix all with a wooden spoon, and after a few minutes begin to add boiling stock gradually; stir with the spoon, so that the rice whilst cooking may absorb the stock; when it is half cooked add a few spoonsful of good gravy and a sweetbread or sheep's brains (previously scalded and cut up in pieces), and, if you like, a little powdered saffron dissolved in a spoonful of stock and three tablespoonsful of grated Parmesan and Cheddar mixed. Stir well until the rice is quite cooked, but take care not to get it into a pulp.

No. 192. Risotto alla Spagnuola

Ingredients: Rice, pork, ham, onions, tomatoes, butter, stock, vegetables, Parmesan.

Put a small bit of onion and an ounce of butter into a saucepan, add half a pound of tomatoes cut up and fry for a few minutes. Then put in some bits of loin of pork cut into dice and some bits of lean ham. After a time add four ounces of rice and good stock, and as soon as it begins to boil put on the cover and put the saucepan on a moderate fire. When the rice is half cooked add any sort of vegetable, by preference peas, asparagus cut up, beans, and cucumber cut up, cook for another quarter of an hour, and serve with grated Parmesan and Cheddar mixed and good gravy.

No. 193. Risotto alla Capuccina

Ingredients: Risotto (No. 190) eggs, truffles, smoked tongue, butter.

Make a good risotto, and when cooked put it into a fireproof dish. When cold cut into shapes with a dariole mould and fry for a few minutes in butter, then turn the darioles out, scoop out a little of each and fill it with eggs beaten up, cover each with a slice of truffle and garnish with a little chopped tongue. Put them in the oven for ten minutes.

No. 194. Risotto alla Parigina

Ingredients: Risotto (No. 190), game, sauce, butter.

Make a good risotto, and when cooked pour it into a fireproof dish, let it get cold, and then cut it out with a dariole mould, or else form it into little balls about the size of a pigeon's egg. Fry these in butter and serve with a rich game sauce poured over them.

No. 195. Ravioli

Ingredients: Flour, eggs, butter, salt, forcemeat, Parmesan, gravy or stock.

Make a paste with a quarter pound of flour, the yolk of two eggs, a little salt and two ounces of butter. Knead this into a firm smooth paste and wrap it up in a damp cloth for half an hour, then roll it out as thin as possible, moisten it with a paste-brush dipped in water, and cut it into circular pieces about three inches in diameter. On each piece put about a teaspoonful of forcemeat of fowl, game, or fish mixed with a little grated Parmesan and the yolks of one or two eggs. Fold the paste over the forcemeat and pinch the edges together, so as to give them the shape of little puffs; let them dry in the larder, then blanch by boiling them in stock for quarter of an hour and drain them in a napkin. Butter a fireproof dish, put in a layer of the ravioli, powder them over with grated Parmesan, then another layer of ravioli and more Parmesan. Then add enough very good gravy to cover them, put the dish in the oven for about twenty-five minutes, and serve in the dish.

No. 196. Ravioli alla Fiorentina

Ingredients: Beetroot, eggs, Parmesan, milk or cream, nutmeg, spices, salt, flour, gravy.

Wash a beetroot and boil it, and when it is sufficiently cooked throw it into cold water for a few minutes, then drain it, chop it up and add to it four eggs, one ounce of grated Parmesan, one ounce of grated Cheddar, two and a half ounces of boiled cream or milk, a small pinch of nutmeg and a little salt. Mix all well together into a smooth firm paste, then roll into balls about the size of a walnut, flour them over well, let them dry for half an hour, then drop them very carefully one by one into boiling stock and when they float on the top take them out with a perforated ladle, put them in a deep dish, dust them over with Parmesan and pour good meat or game gravy over them.

No. 197. Gnocchi alla Romana

Ingredients: Semolina, butter, Parmesan, eggs, nutmeg, milk, cream.

Boil half a pint of milk in a saucepan, then add two ounces of butter, four ounces of semolina, two tablespoonsful of grated Parmesan, the yolks of three eggs, and a tiny pinch of nutmeg. Mix all well together, then let it cool, and spread out the paste so that it is about the thickness of a finger. Put a little butter and grated Parmesan and two tablespoonsful of cream in a fireproof dish, cut out the semolina paste with a small dariole mould and put it in the dish. Dust a little more Parmesan over it, put it in the oven for five minutes and serve in the dish.

No. 198. Gnocchi alla Lombarda

Ingredients: Potatoes, flour, salt, Parmesan and Gruyere cheese, butter, milk, eggs.

Boil two or three big potatoes, and pass them through a hair sieve, mix in two tablespoonsful of flour, an egg beaten up, and enough milk to form a rather firm paste; stir until it is quite smooth. Roll it into the shape of a German sausage, cut it into rounds about three quarters of an inch thick, and put it into the larder to dry for about half an hour. Then drop the gnocchi one by one into boiling salted water and boil for ten minutes. Take them out with a slice, and put them in a well-buttered fireproof dish, add butter between each layer, and strew plenty of grated Parmesan and Cheddar over them. Put them in the oven for ten minutes, brown the top with a salamander, and serve very hot.

No. 199. Frittata di Riso (Savoury Rice Pancake)

Ingredients: Rice, milk, salt, butter, cinnamon, eggs, Parmesan.

Boil quarter of a pound of rice in milk until it is quite soft and pulpy, drain off the milk and add to the rice an ounce of butter, two tablespoonsful of grated Parmesan, and a pinch of cinnamon, and when it has got rather cold, the yolks of four eggs beaten up. Mix all well together, and with this make a pancake with butter in a frying pan.

Omelettes And Other Egg Dishes

No. 200. Uova al Tartufi (Eggs with Truffles)

Ingredients: Eggs, butter, cream, truffles, Velute sauce, croutons.

Beat up six eggs, pass them through a sieve, and put them into a saucepan with two ounces of butter and two tablespoonsful of cream. Put the saucepan in a bain-marie, and stir so that the eggs may not adhere. Sautez some slices of truffle in butter, cover them with Velute sauce (No. 2) and a glass of Marsala, and add them to the eggs. Serve very hot with fried and glazed croutons. Instead of truffles you can use asparagus tips, peas, or cooked ham.

No. 201. Uova al Pomidoro (Eggs and Tomatoes)

Ingredients: Eggs, salt, tomatoes, onion, parsley, butter, pepper.

Cut up three or four tomatoes, and put them into a stewpan with a piece of butter the size of a walnut and a clove of garlic with a cut in it. Put the lid on the stewpan and cook till quite soft, then take out the garlic, strain the tomatoes through a fine strainer into a bain-marie, beat up two eggs and add them to the tomatoes, and stir till quite thick, then put in two tablespoonsful of grated cheese, and serve on toast.

No. 202. Uova ripiene (Canapes of Egg)

Ingredients: Eggs, butter, salt, pepper, nutmeg, cheese, parsley, mushrooms, Bechamel and Espagnole sauce, stock.

Boil as many eggs as you want hard, and cut them in half lengthwise; take out the yolks and mix them with some fresh butter, salt, pepper, very little nutmeg, grated cheese, a little chopped parsley, and cooked mushrooms also chopped. Then mix two tablespoonsful of good Bechamel sauce (No. 3) with the raw yolk of one or two eggs and add it to the rest. Put all in a saucepan with an ounce of butter and good stock, then fill up the white halves with the mixture, giving them a good shape; heat them in a bain-marie, and serve with a very good clear Espagnole sauce (No. 1).

No. 203. Uova alla Fiorentina (Eggs)

Ingredients: Eggs, butter, Parmesan, cream, flour, salt, pepper, curds.

Boil as many eggs as you require hard, then cut them in half and take out the yolks and pound them in a mortar with equal quantities of butter and curds, a tablespoonful of grated Parmesan, salt and pepper. Put this in a saucepan and add the yolks of eight eggs and the white of one (this is for twelve people), mix all well together and reduce a little. With this mixture fill the hard whites of the eggs and spread the rest of the sauce on the bottom of the dish, and on this place the whites. Then in another saucepan mix half a gill of cream and an ounce of butter, a dessert-spoonful of flour, salt, and pepper; let this boil for a minute, and then glaze over the eggs in the dish with it, and on the top of each egg put a little bit of butter, and over all a powdering of grated cheese. Put this in the oven, pass the salamander over the top, and when the cheese is coloured serve at once.

No. 204. Uova in fili (Egg Canapes)

Ingredients: Eggs, butter, mushrooms, onions, flour, white wine, fish or meat stock, salt, pepper, croutons of bread.

Put into a saucepan two ounces of butter, three large fresh mushrooms cut into slices, and an onion cut up, fry them slightly, and when the onion begins to colour add a spoonful of flour, a quarter of a glass of Chablis, salt and pepper, and occasionally add a spoonful of either fish or meat stock. Let this simmer for half an hour, so as to reduce it to a thick sauce. Then boil as many eggs as you want hard; take out the yolks, but keep them whole. Cut up the whites into slices, and add them to the above sauce, pour the sauce into a dish, and on the top of it place the whole yolks of egg, each on a crouton of bread.

No. 205. Frittata di funghi (Mushroom Omelette)

Ingredients: Mushrooms, butter, eggs, bread crumbs, Parmesan, marjoram, garlic.

Clean four or five mushrooms, cut them up, and put them into a frying-pan with one and a half ounces of butter, a clove of garlic with two cuts in it, and a little salt; fry them lightly till the mushrooms are nearly cooked, and then take out the garlic. In the meantime beat up separately the yolks and the whites of two or three eggs, add a little crumb of bread soaked in water, a tablespoonful of grated Parmesan, and two leaves of marjoram; go on beating all up until the crumb of bread has become entirely absorbed by the eggs, then pour this mixture into the frying-pan with the mushrooms, mix all well together and make an omelette in the usual way.

No. 206. Frittata con Pomidoro (Tomato Omelette)

Ingredients: Eggs, tomatoes, butter, marjoram, parsley, spice.

Peel two tomatoes and take out the seeds; then mix them with an ounce of butter, chopped marjoram, parsley, and a tiny pinch of spice. Add three eggs beaten up (the yolks and whites separately), and make an omelette.

No. 207. Frittata con Asparagi (Asparagus Omelette)

Ingredients: Eggs, asparagus, butter, ham, herbs, cheese.

Blanch a dozen heads of asparagus and cook them slightly, then cut them up and mix with two ounces of butter, bits of cut-up ham, herbs, and a tablespoonful of grated Parmesan. Add them to three beaten-up eggs and make an omelette.

No. 208. Frittata con erbe (Omelette with Herbs)

Ingredients: Eggs, onions, sorrel, mint, parsley, asparagus, marjoram, salt, pepper, butter.

Chop a little sorrel, a small bit of onion, mint, parsley, marjoram, and fry in two ounces of butter, add some cut-up asparagus, salt, and pepper. Then add three eggs beaten up and a little grated cheese, and make your omelette.

No. 209. Frittata Montata (Omelette Souffle)

Ingredients: Eggs, Parmesan, pepper, parsley.

Beat up the whites of three eggs to a froth and the yolks separately with a tablespoonful of grated Parmesan, chopped parsley, and a little pepper. Then mix them and make a light omelette.

No. 210. Frittata di Prosciutto (Ham Omelette)

Ingredients: Eggs, ham, Parmesan, mint, pepper, clotted cream.

Beat up three eggs and add to them two tablespoonsful of clotted cream, one tablespoonful of chopped ham, one of grated Parmesan, chopped mint and a little pepper, and make the omelette in the usual way.

Sweets and Cakes

No. 211. Bodino of Semolina

Ingredients: Semolina, milk, eggs, castor sugar, lemon, sultanas, rum, butter, cream, or Zabajone (No. 222).

Boil one and a half pints of milk with four ounces of castor sugar, and gradually add five ounces of semolina, boil for a quarter of an hour more and stir continually with a wooden spoon, then take the saucepan off the fire, and when it is cooled a little, add the yolks of six and the whites of two eggs well beaten up, a little grated lemon peel, three-quarters of an ounce of sultanas and two small glasses of rum. Mix well, so as to get it very smooth, pour it into a buttered mould and serve either hot or cold. If cold, put whipped cream flavoured with stick vanilla round the dish; if hot, a Zabajone (No. 222).

No. 212. Crema rappresa (Coffee Cream)

Ingredients: Coffee, cream, eggs, sugar, butter.

Bruise five ounces of freshly roasted Mocha coffee, and add it to three-quarters of a pint of boiling cream; cover the saucepan, let it simmer for twenty minutes, then pass through a bit of fine muslin.

In the meantime mix the yolks of ten eggs and two whole eggs with eight ounces of castor sugar and a glass of cream; add the coffee cream to this and pass the whole through a fine sieve into a buttered mould. Steam in a bain-marie for rather more than an hour, but do not let the water boil; then put the cream on ice for about an hour, and before serving turn it out on a dish and pour some cream flavoured with stick vanilla round it.

No. 213. Crema Montata alle Fragole (Strawberry Cream)

Ingredients: Cream, castor sugar, Maraschino, strawberries or strawberry jam.

Put a pint of cream on ice, and after two hours whip it up. Pass three tablespoonsful of strawberry jam through a sieve and add two tablespoonsful of Maraschino; mix this with the cream and build it up into a pyramid. Garnish with meringue biscuits and serve quickly. You may use fresh strawberries when in season, but then add castor sugar to taste.

No. 214. Croccante di Mandorle (Cream Nougat)

Ingredients: Almonds, sugar, lemon juice, butter, castor sugar, pistachios, preserved fruits.

Blanch half a pound of almonds, cut them into shreds and dry them in a slow oven until they are a light brown colour; then put a quarter pound of lump sugar into a saucepan and caramel it lightly; stir well with a wooden spoon. When the sugar is dissolved, throw the hot almonds into it and also a little lemon juice. Take the saucepan off the fire and mix the almonds with the sugar, pour it into a buttered mould and press it against the sides of the mould with a lemon, but remember that the casing of sugar must be very thin. (You may, if you like, spread out the mixture on a flat dish and line the mould with your hands, but the sugar must be kept hot.) Then take it out of the mould and decorate it with castor sugar, pistacchio nuts, and preserved fruits. Fill this case with whipped cream and preserved fruits or fresh strawberries.

No. 215. Crema tartara alla Caramella (Caramel Cream)

Ingredients: Cream, eggs, caramel sugar, vanilla or lemon flavouring.

Boil a pint of cream and give it any flavour you like. When cold, add the yolks of eight eggs and two tablespoonsful of castor sugar, mix well and pass it through a sieve; then burn some sugar to a caramel, line a smooth mould with it and pour the cream into it. Boil in a bain-marie for an hour and serve hot or cold.

No. 216. Cremona Cake

Ingredients: Ground rice, ground maize, sugar, one orange, eggs, salt, cream, Maraschino, almonds, preserved cherries.

Weigh three eggs, and take equal quantities of castor sugar, butter, ground rice and maize (the last two together); make a light paste with them, but only use one whole egg and the yolks of the two others, add the scraped peel of an orange and a pinch of salt. Roll this paste out to the thickness of a five-shilling piece, colour it with the yolk of an egg and bake it in a cake tin in a hot oven until it is a good colour, then take it out and cut it into four equal circular pieces. Have ready some well-whipped cream and flavour it with Maraschino, put a thick layer of this on one of the rounds of pastry, then cover it with the next round, on which also put a layer of cream, and so on until you come to the last round, which forms the top of the cake. Then split some almonds and colour them in the oven, cover the top of the cake with icing sugar flavoured with orange, and decorate the top with the almonds and preserved cherries.

No. 217. Cake alla Tolentina

Ingredients: Sponge-cake, jam, brandy or Maraschino, cream, pine-apple.

Make a medium-sized sponge-cake; when cold cut off the top and scoop out all the middle and leave only the brown case; cover the outside with a good coating of jam or red currant jelly, and decorate it with some of the white of the cake cut into fancy shapes. Soak the rest of the crumb in brandy or Maraschino and mix it with quarter of a pint of whipped cream and bits of pineapple cut into small dice; fill the cake with this; pile it up high in the centre and decorate the top with the brown top cut into fancy shapes.

No. 218. Riso all'Imperatrice

Ingredients: Rice, sugar, milk, ice, preserved fruits, blanc-mange, Maraschino, cream.

Boil two dessert-spoonful of rice and one of sugar in milk. When sufficiently boiled, drain the rice and let it get cold. In the meantime place a mould on ice, and decorate it with slices of preserved fruit, and fix them to the mould with just enough nearly cold dissolved isinglass to keep them in place. Also put half a pint of blanc-mange on the ice, and stir it till it is the right consistency, gradually add the boiled rice, half a glass of Maraschino, some bits of pineapple cut in dice, and last of all half a pint of whipped cream. Fill the mould with this, and when it is sufficiently cold, turn it out and serve with a garnish of glace fruits or a few brandy cherries.

No. 219. Amaretti leggieri (Almond Cakes)

Ingredients: Almonds (sweet and bitter), eggs, castor sugar.

Blanch equal quantities of sweet and bitter almonds, and dry them a little in the oven, then pound them in a mortar, and add nearly double their quantity of castor sugar. Mix with the white of an egg well beaten up into a snow, and shape into little balls about the size of a pigeon's egg. Put them on a piece of stout white paper, and bake them in a very slow oven. They should be very light and delicate in flavour.

No. 220. Cakes alla Livornese

Ingredients: Almonds, eggs, sugar, salt, potato flour, butter.

Pound two ounces of almonds, and mix them with the yolks of two eggs and a spoonful of castor sugar flavoured with orange juice. Then mix two ounces of sugar with an egg, and to this add the almonds, a pinch of salt, and gradually strew in one and a half ounces of potato flour. When it is all well mixed, add one ounce of melted butter, shape the cakes and bake them in a slow oven.

No. 221. Genoese Pastry

Ingredients: Eggs, sugar, butter, flour, almonds, orange or lemon, brandy.

Weigh four eggs, and take equal weights of castor sugar, butter, and flour. Pound three ounces of almonds, and mix them with an egg, melt the butter, and mix all the ingredients with a wooden spoon in a pudding basin for ten minutes, then add a little scraped orange or lemon peel, and a dessert-spoonful of brandy. Spread out the paste in thin layers on a copper baking sheet, cover them with buttered paper, and bake in a moderately hot oven.

These cakes must be cut into shapes when they are hot, as otherwise they will break.

No. 222. Zabajone

Ingredients: Eggs, sugar, Marsala, Maraschino or other light-coloured liqueur, sponge fingers.

Zabajone is a kind of syllabub. It is made with Marsala and Maraschino, or Marsala and yellow Chartreuse. Reckon the quantities as follows: for each person the yolks of three eggs, one teaspoonful of castor sugar to each egg, and a wine-glass of wine and liqueur mixed. Whip up the yolks of the eggs with the sugar, then gradually add the wine. Put this in a bain-marie, and stir until it has thickened to the consistency of a custard. Take care, however, that it does not boil. Serve hot in custard glasses, and hand sponge fingers with it.

No. 223. Iced Zabajone

Ingredients: Eggs, castor sugar, Marsala, cinnamon, lemon, stick vanilla, rum, Maraschino, butter, ice.

Mix the yolks of ten eggs, two dessert-spoonful of castor sugar, and three wine-glasses of Marsala, add half a stick of vanilla, a small bit of whole cinnamon, and the peel of half a lemon cut into slices.

Whip this up lightly over a slow fire until it is nearly boiling and slightly frothy; then remove it, take out the cinnamon, vanilla, and lemon pool, and whip up the rest for a minute or two away from the fire. Add a tablespoonful of Maraschino and one of rum, and, if you like, a small quantity of dissolved isinglass. Stir up the whole, pour it into a silver souffle dish, and put it on ice. Serve with sponge cakes or iced wafers.

No. 224. Pan-forte di Siena (Sienese Hardbake)

Ingredients: Honey, almonds, filberts, candied lemon peel, pepper, cinnamon, chocolate, corn flour, large wafers.

Boil half a pound of honey in a copper vessel, and then add to it a few blanched almonds and filberts cut in halves or quarters and slightly browned, a little candied lemon peel, a dust of pepper and powdered cinnamon and a quarter pound of grated chocolate. Mix all well together, and gradually add a tablespoonful of corn flour end two of ground almonds to thicken it. Then take the vessel off the fire, spread the mixture on large wafers, and make each cake about an inch thick. Garnish them on the top with almonds cut in half, and dust over a little powdered sugar and cinnamon, then put them in a very slow oven for an hour.

NEW CENTURY SAUCE * * The New Century Sauce may be bought at Messrs. Lazenby's, Wigmore Street, W

No. 225. Fish Sauce

Add one dessert-spoonful of the sauce to a quarter pint of melted butter sauce.

No. 226. Sauce Piquante (for Meat, Fowl, Game, Rabbit, &c.)

One dessert-spoonful to a quarter pint of ordinary brown or white stock. It may be thickened by a roux made by frying two ounces of butter with two ounces of flour.

No. 227. Sauce for Venison, Hare, &c.

Two dessert-spoonsful of New Century Sauce to half a pint of game gravy or sauce, and a small teaspoonful of red currant jelly.

No. 228. Tomato Sauce Piquante

Fry three medium-sized tomatoes in one and a half ounce of butter. Pass this through a sieve, then boil it up in a bain-marie till it thickens, and add one dessertspoonful of New Century Sauce.

No. 229. Sauce for Roast Pork, Ham, &c.

Add to any ordinary white or brown sauce one dessert-spoonful of New Century Sauce and two of port or Burgundy if the sauce is brown, two of Chablis if white.

No. 230. For masking Cutlets, &c.

Making a roux by frying two ounces of butter with two ounces of flour, and add two tablespoonsful of boiling stock. Stir in one dessert-spoonful of New Century Sauce. Let it get cold, and it will then be quite firm and ready for masking cutlets, &c.

www.ingramcontent.com/pod-product-compliance
Lightning Source LLC
Chambersburg PA
CBHW031631210526
45464CB00004B/1840